BIG BOOK OF
PLANT AND FLOWER
ILLUSTRATIONS

Selected and Arranged by
MAGGIE KATE

DOVER PUBLICATIONS, INC.
Mineola, New York

Bibliographical Note

Big Book of Plant and Flower Illustrations is a new work, first published by Dover Publications, Inc., in 2000.

DOVER *Pictorial Archive* SERIES

International Standard Book Number: 0-486-40946-5

Manufactured in the United States of America
Dover Publications, Inc., 31 East 2nd Street, Mineola, N.Y. 11501

Publisher's Note

The *Big Book of Plant and Flower Illustrations* offers a generous sampling of the work of eleven outstanding nature artists. Included in this compendium are authentic views of hundreds of common and rare species of plants and flowers from all over the world, and from a multitude of diverse climates and habitats. The index beginning on page 119 provides easy access to the page locations of specific entries, and those interested in more in-depth coverage of any of the individual categories may refer to "Sources of the Illustrations" on page 124.

The book opens with "Garden Flowers." During the spring and summer months most of these plants are easily located in nearby gardens. And the best guide to coloring the plates are the plants themselves. Next is "Wildflowers," which includes favorites and frequently met flowers, with a few unusual items that are specially suited for coloring. Here are some of the natural glories of the American landscape, flowers that make their homes in swamps, thickets, rock clefts—almost anywhere roots can take hold. "House Plants" give endless pleasure by bringing a touch of nature into indoor environments. Unlike other home furnishings and decorative accessories, plants do not remain static for long periods of time, altering in appearance from season to season.

Since ancient times, "Herbs" have been grown for their culinary and medicinal value. The monasteries of the Middle Ages were never without their herb gardens, and the great Elizabethan herbals contained fascinating lore about the uses of thyme, savory, rosemary, and basil. In recent years there has been a revival of interest in these unusual plants that provide fragrance, flavoring, medicine, and nourishment. Closely related to this topic is "Medicinal Plants." With the recent widespread destruction of tropical rain forests has come the anguished cry for their preservation. Not the least important reason is that so many plants of medicinal value, actual or potential, grow there. Even with the modern ability to synthesize drugs, wild plants are ever of crucial importance in pharmacology. "Common Weeds" are often interesting plants, noteworthy because of their high vigor, adaptability, and fecundity. Some weeds are attractive, while others have curious properties. In Renaissance Europe their presence was used as a guide to mineral deposits since they may indicate the chemistry of the soil in which they grow. Other weeds are medicinal, some are narcotic, and some are edible. The weeds in this book are the important land weeds that are practically universal in the U.S.

"Cacti" are among the most remarkable and most misunderstood plants in the world. Although some species are capable of surviving in deserts, others flourish in habitats ranging from tropical jungles to windswept mountainsides. Cacti are native to the Americas, but cactus plants and seeds have been spread around the world since ancient times, and cacti may today be found growing almost anywhere. Cactus flowers are among the most beautiful flowers in the world, and for this book, cacti which are especially unusual or attractive have been chosen.

Nature has lavished some of her boldest shapes and colors on the blossoms of plants that grow in tropical regions. "Tropical Flowers" includes exciting specimens from such diverse areas as Mexico, Central and South America, Africa, southern Asia and the island world of the Indian and Pacific Oceans. Blossoms occur on trees, shrubs, herb, or vining plants. "Hawaiian Plants" and animals were thriving on this chain of islands in the Pacific Ocean for at least a million years before the first Polynesians arrived. It is estimated that over 2,000 species of endemic plants and about 60 varieties of endemic birds existed on these islands when the first Polynesians arrived there more than 1,500 years ago.

"Orchids" represent the largest family of flowering plants in terms of number of species. At one time, these exotic plants were grown only in the greenhouses of the wealthy. Now, as a result of the wide availability of commercially grown plants, orchids are far more common in amateur greenhouses and home gardens. "Roses," in all their splendid diversity, are the world's best known and most loved flowers. Cultivated since antiquity, they have developed an astounding variety of form, color, scent, and behavior. "Redouté Flowers" includes carefully rendered flower drawings after the paintings of Pierre-Joseph Redouté (1759–1840), best known for his breathtaking portraits of roses. He created many of his works while employed by the Empress Josephine. From simple nosegays of daisies and tulips to more ornate arrangements of orchids, lilies and roses, "Floral Bouquets" evoke the charm and grace of fresh flowers.

"Mushrooms" can be breathtakingly beautiful or undeniably repulsive. Although mushrooms are normally thought of as ephemeral phenomena, the fungus that produces them is usually perennial, living for as long as 400 years while fruiting (forming mushrooms) periodically. Unlike a green plant, it cannot make its own food, and must therefore feed on dead, decaying, or occasionally, living matter. Several thousand different kinds of mushrooms are known in Europe and temperate North America, with some of the more colorful, distinctive and interesting species depicted here. Since the northeastern U.S. has ample moisture and fertile soil, it boasts a rich variety of trees. "Trees of the Northeast" includes several that are illustrated along with distinctive features, such as their fruits and seeds, leaves, barks and flowers.

Finally, the "Miscellaneous" section is a potpourri of plants and flowers encompassing a selection of state flowers, as well as a sampling of plants from several nature preserves and botanical gardens.

Contents

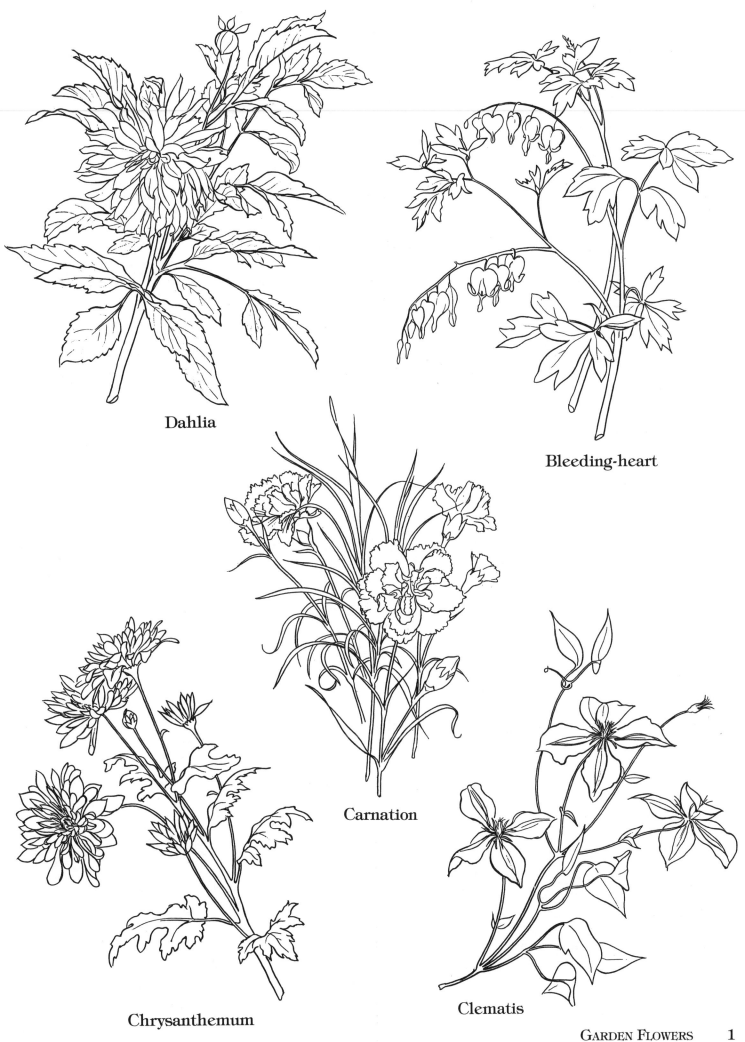

Dahlia

Bleeding-heart

Carnation

Chrysanthemum

Clematis

Coreopsis

Erigeron

Lupine

Sweet William

Cosmos

Daylily

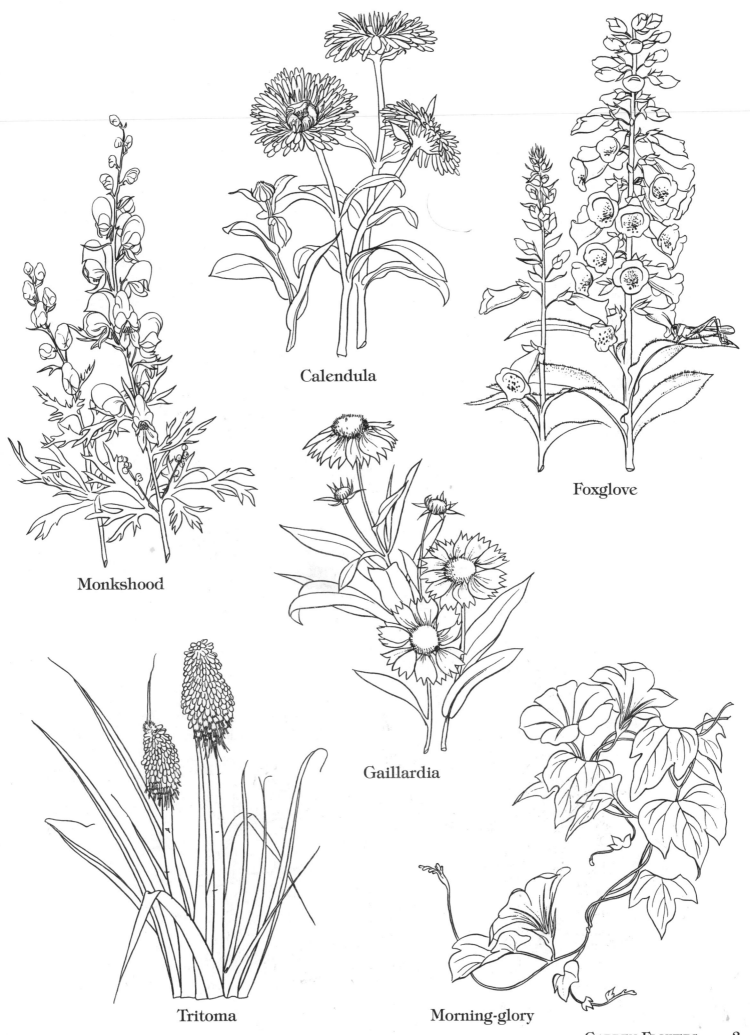

Calendula

Foxglove

Monkshood

Gaillardia

Tritoma

Morning-glory

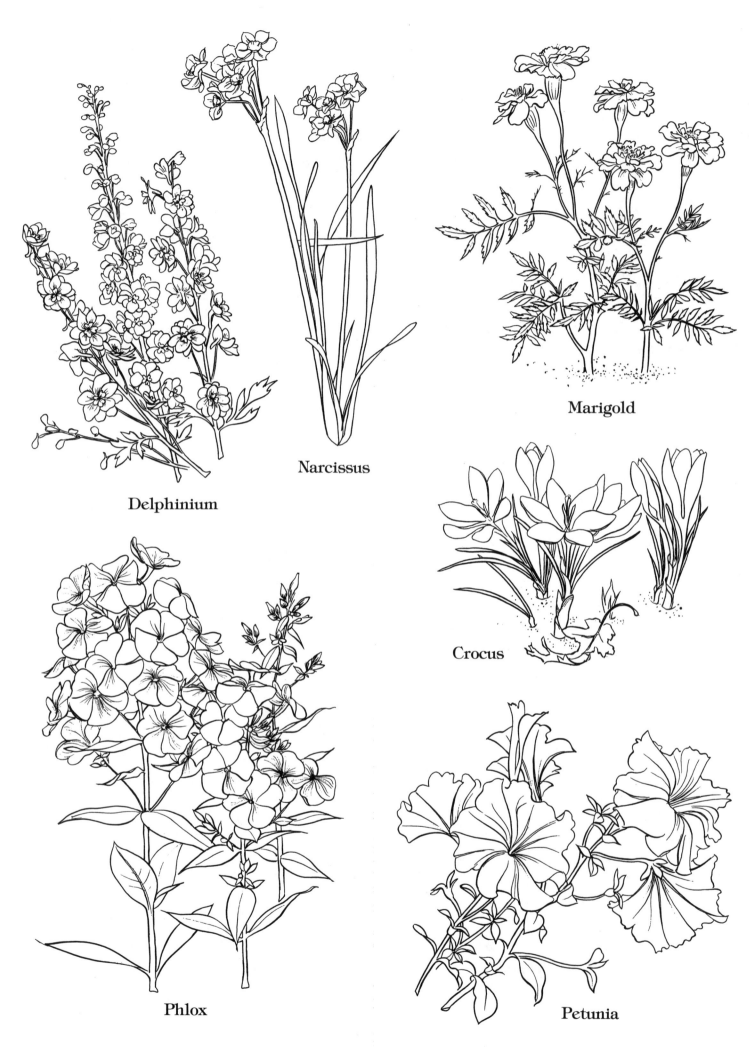

Narcissus

Delphinium

Marigold

Crocus

Phlox

Petunia

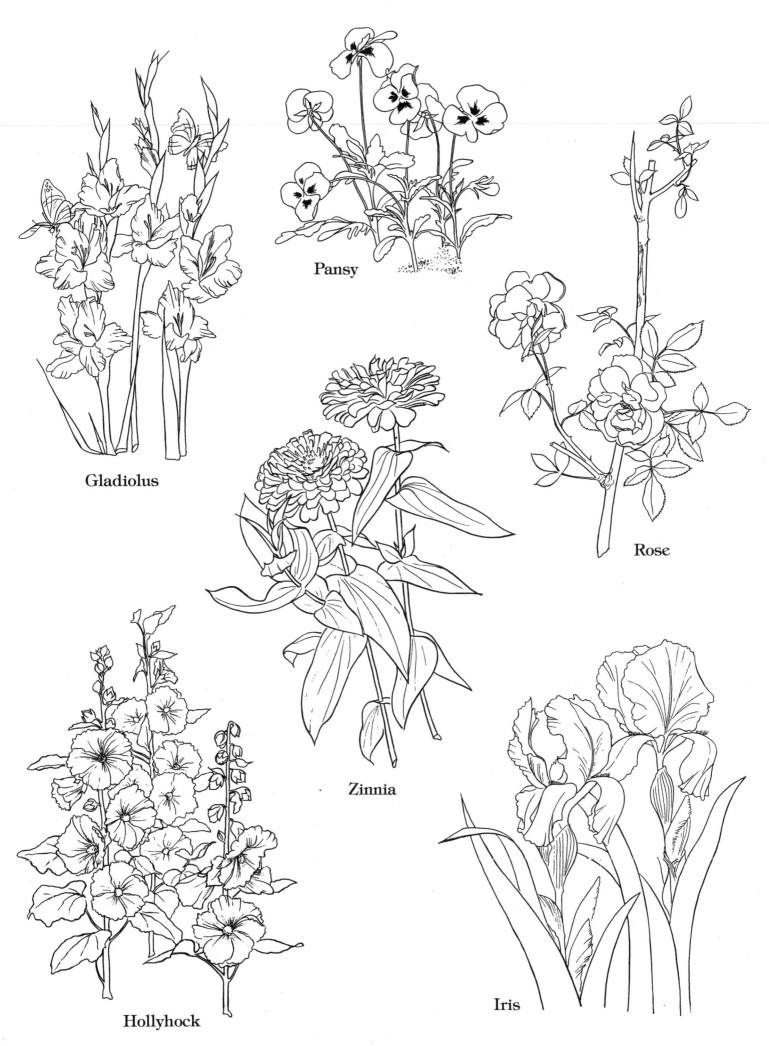

Gladiolus

Pansy

Rose

Zinnia

Hollyhock

Iris

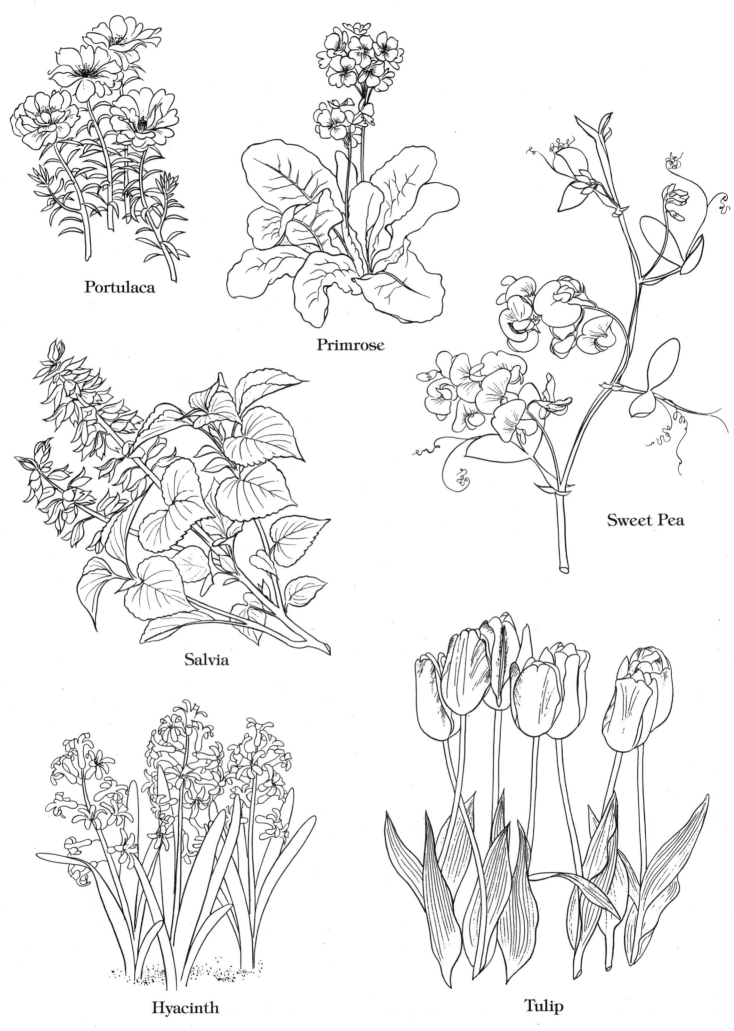

Portulaca

Primrose

Sweet Pea

Salvia

Hyacinth

Tulip

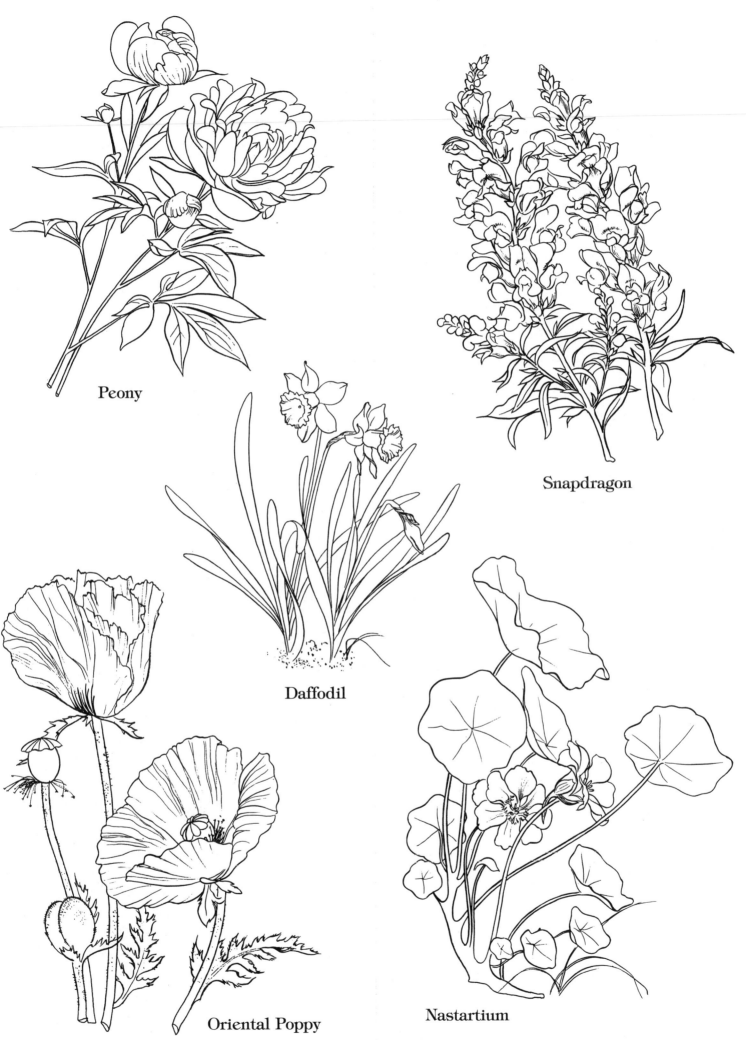

Peony

Snapdragon

Daffodil

Oriental Poppy

Nastartium

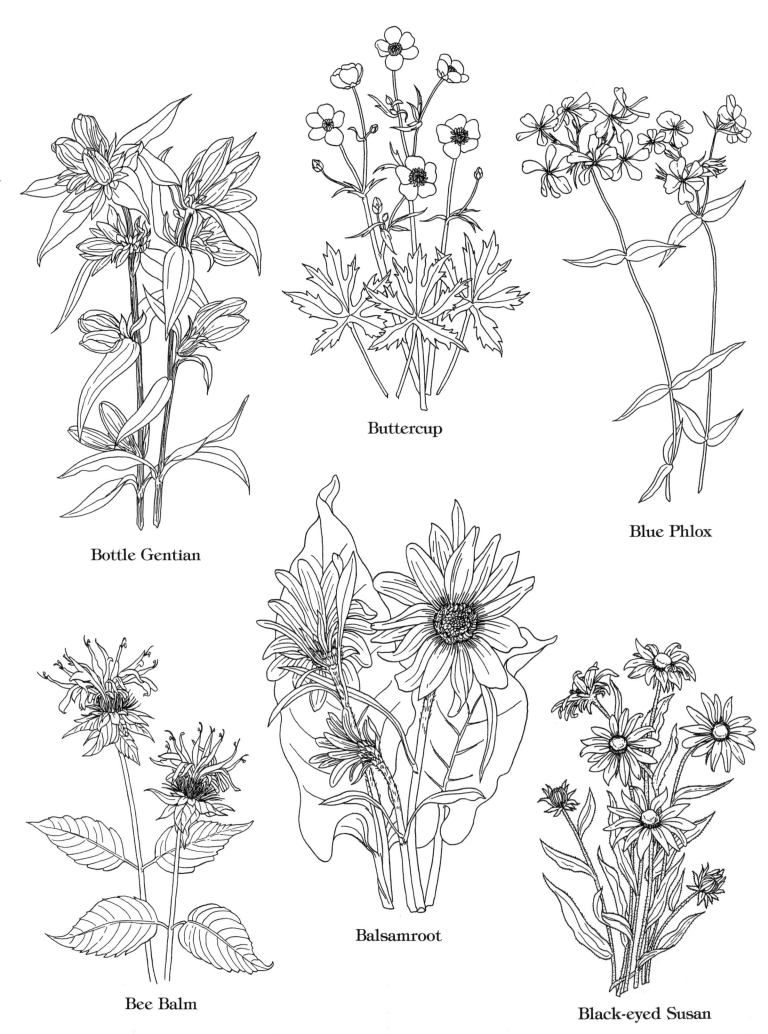

Buttercup

Blue Phlox

Bottle Gentian

Balsamroot

Bee Balm

Black-eyed Susan

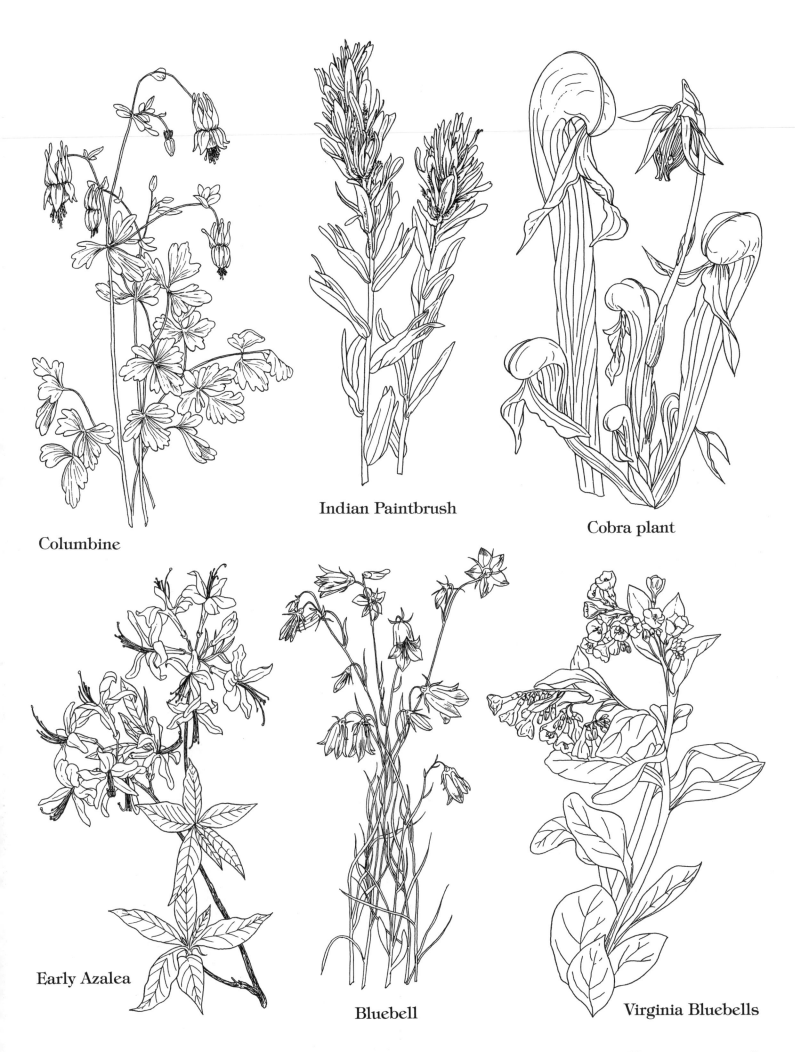

Columbine

Indian Paintbrush

Cobra plant

Early Azalea

Bluebell

Virginia Bluebells

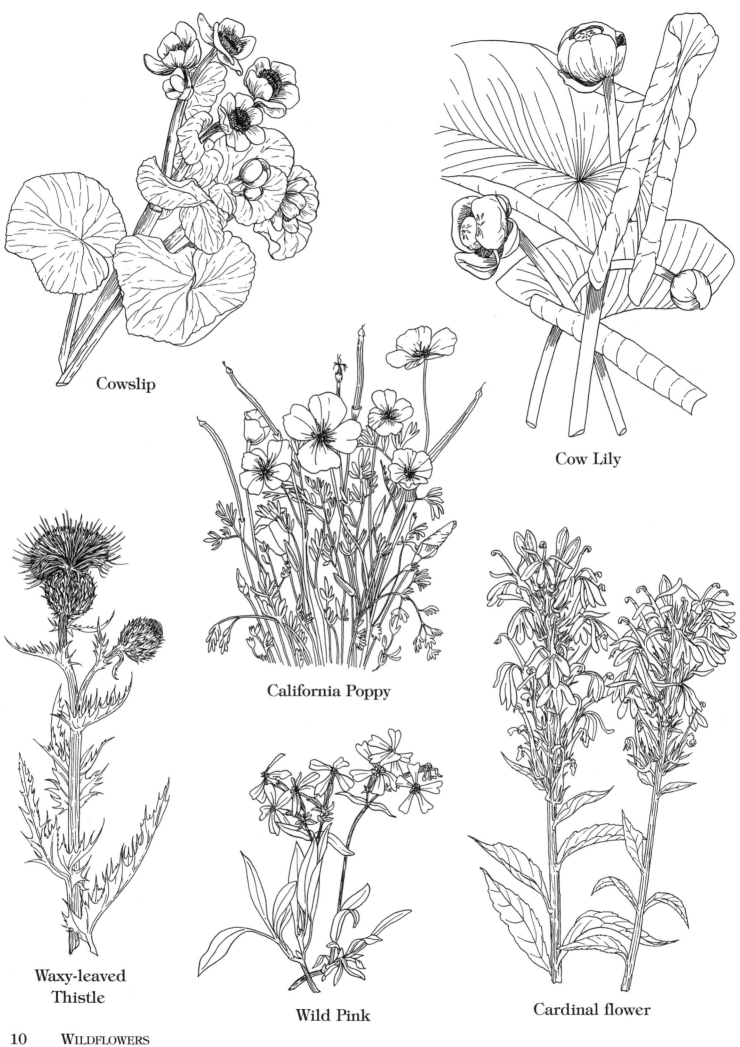

Cowslip

Cow Lily

California Poppy

Waxy-leaved
Thistle

Wild Pink

Cardinal flower

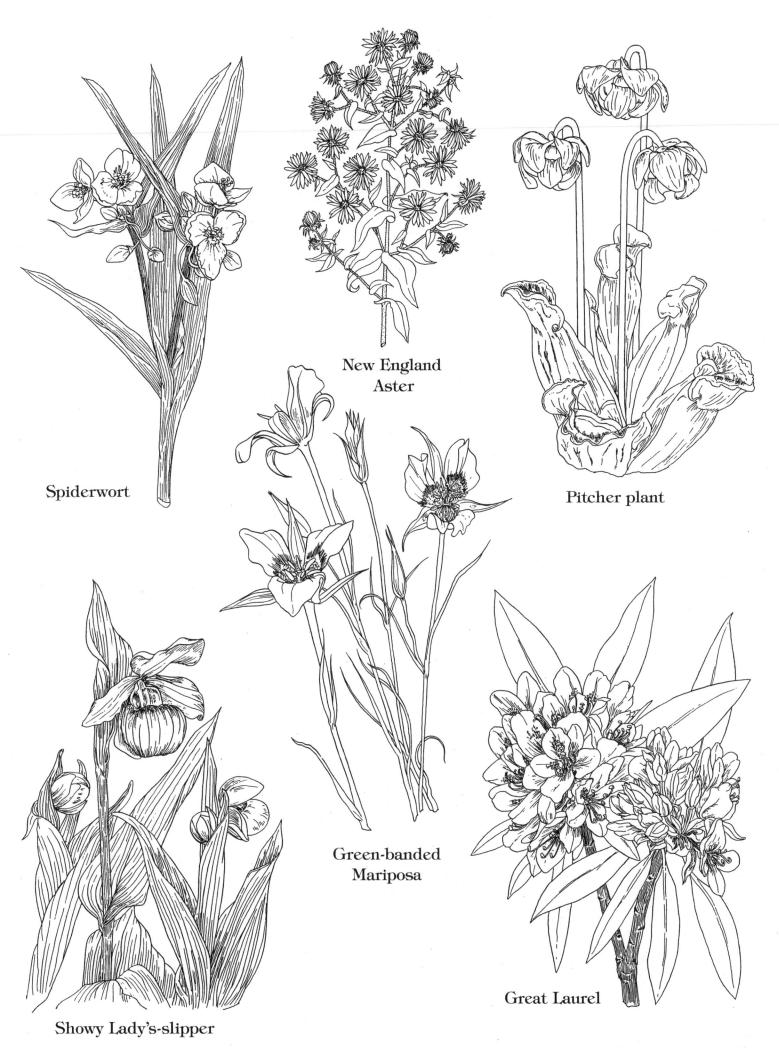

Spiderwort

New England
Aster

Pitcher plant

Green-banded
Mariposa

Showy Lady's-slipper

Great Laurel

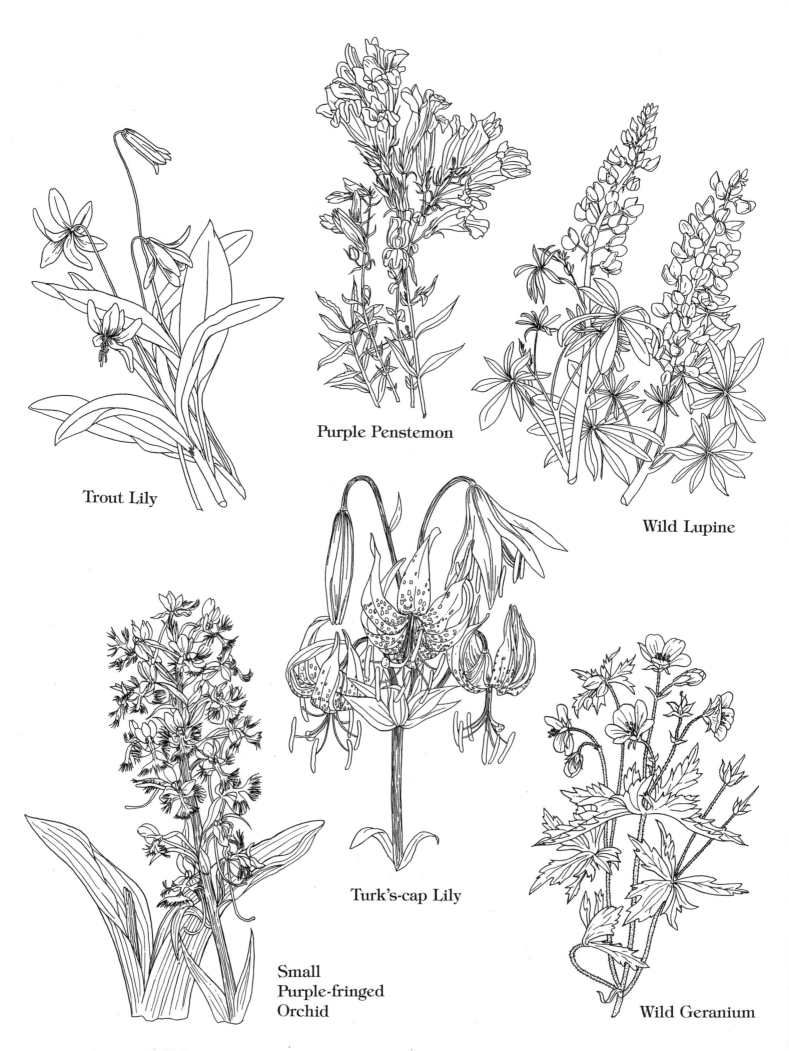

Purple Penstemon

Trout Lily

Wild Lupine

Small
Purple-fringed
Orchid

Turk's-cap Lily

Wild Geranium

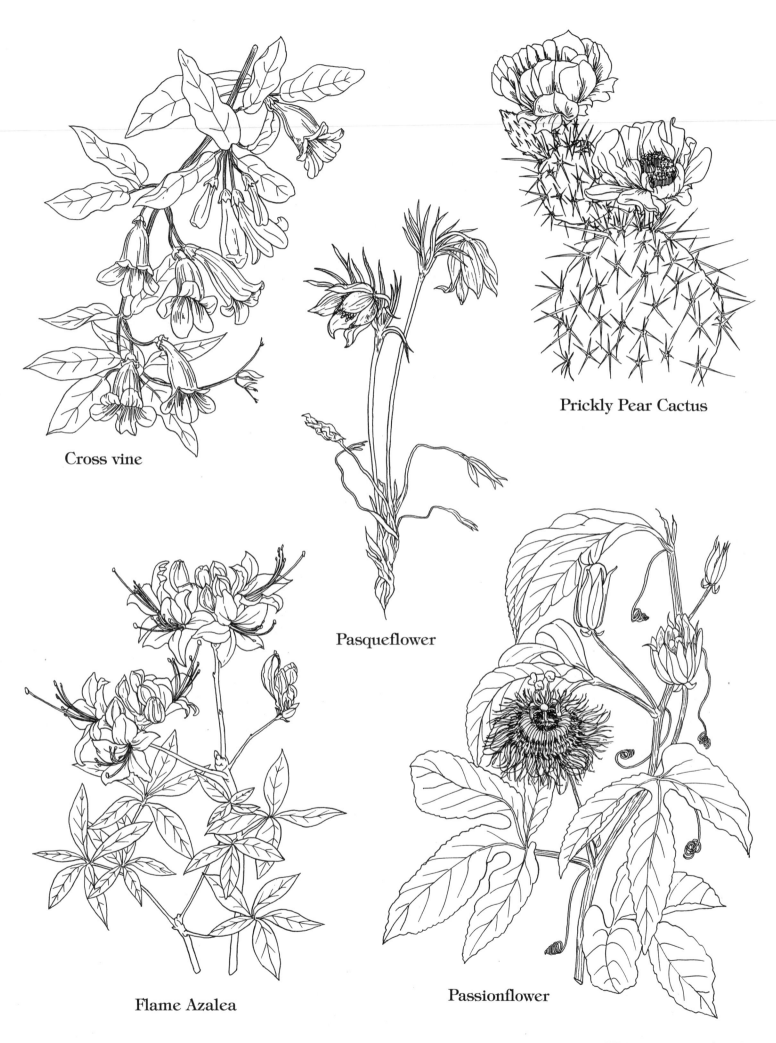

Cross vine

Pasqueflower

Prickly Pear Cactus

Flame Azalea

Passionflower

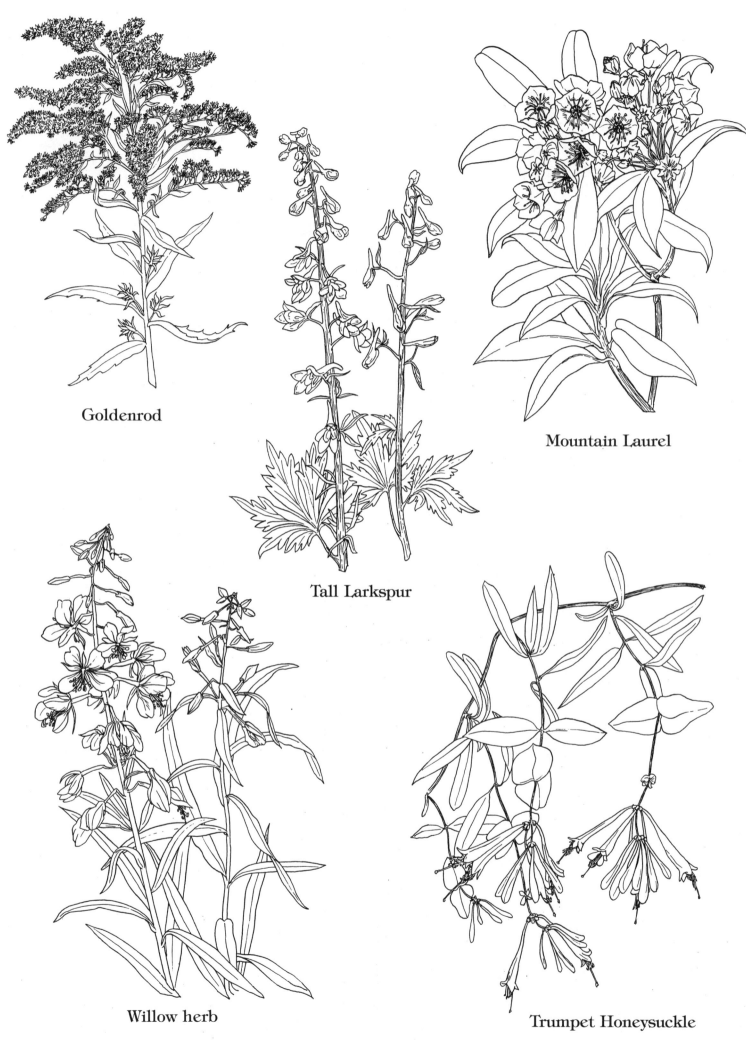

Goldenrod

Tall Larkspur

Mountain Laurel

Willow herb

Trumpet Honeysuckle

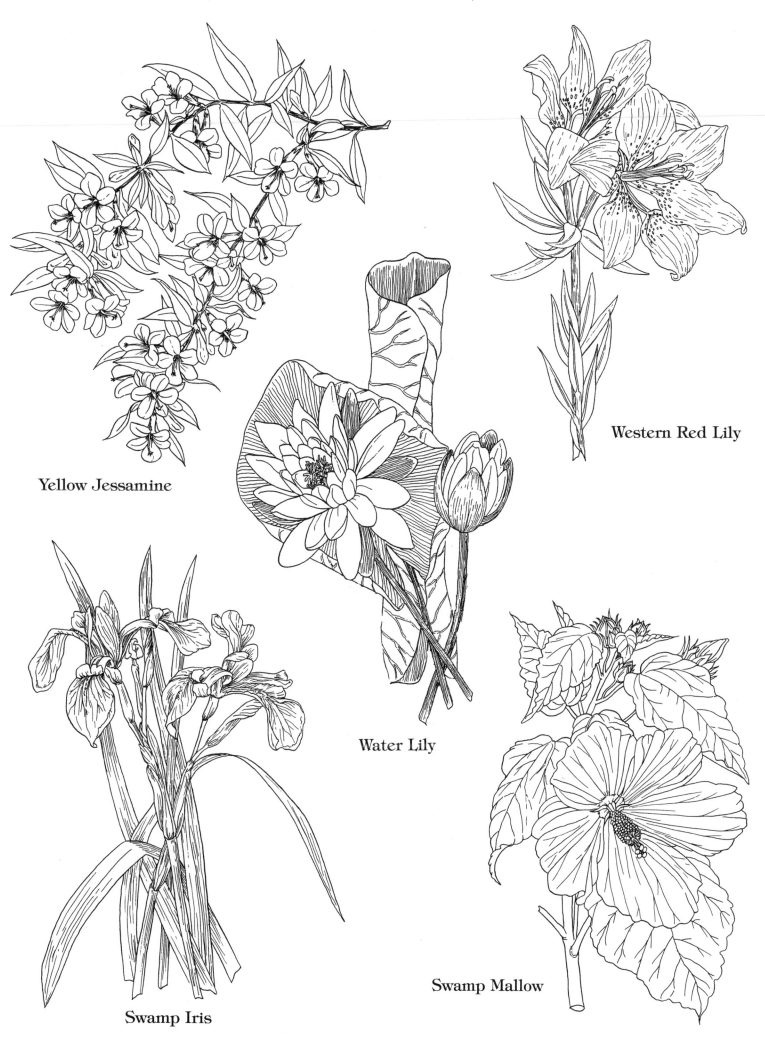

Yellow Jessamine

Western Red Lily

Water Lily

Swamp Iris

Swamp Mallow

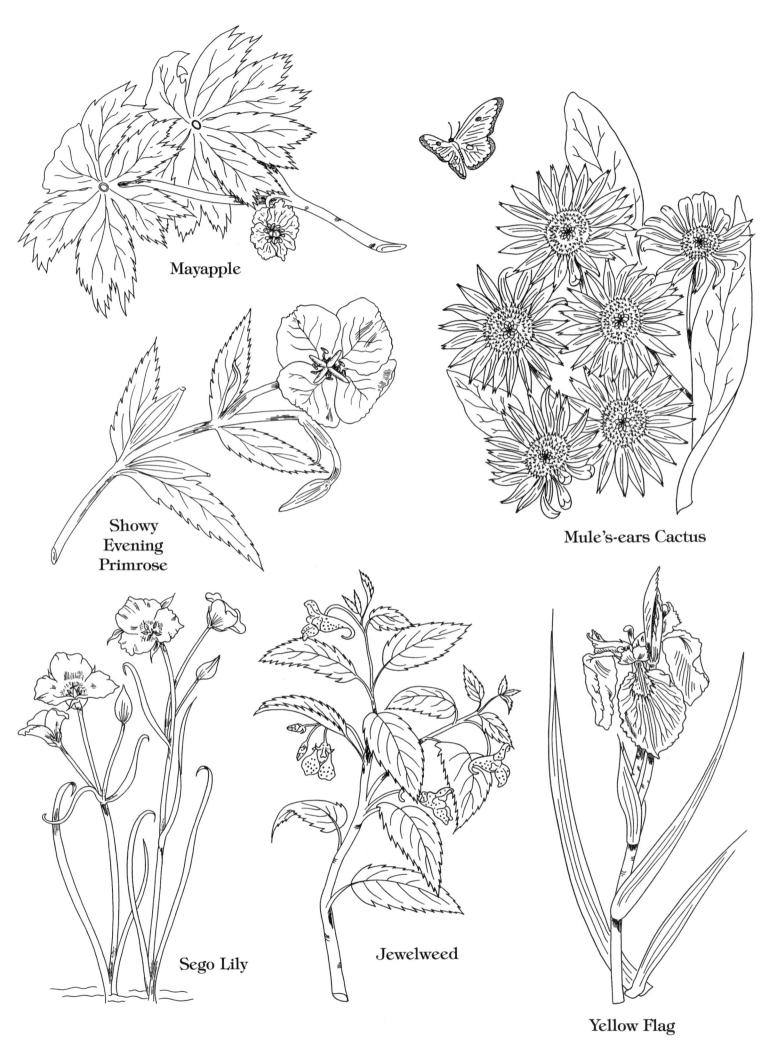

Mayapple

Showy
Evening
Primrose

Mule's-ears Cactus

Sego Lily

Jewelweed

Yellow Flag

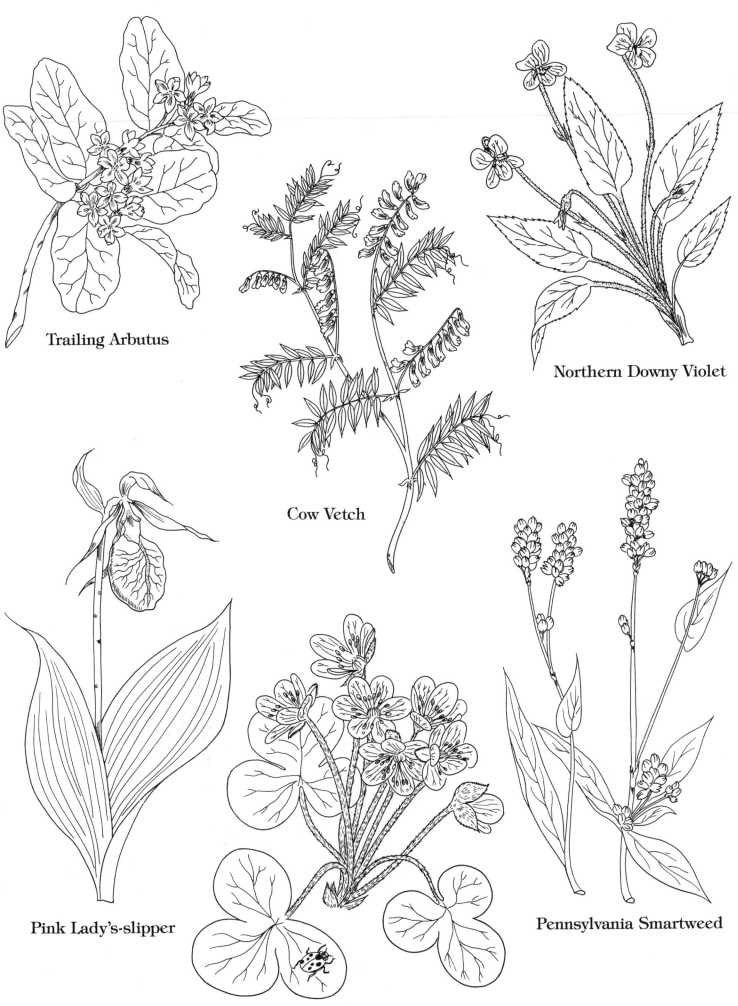

Trailing Arbutus

Cow Vetch

Northern Downy Violet

Pink Lady's-slipper

Round-lobed Hepatica

Pennsylvania Smartweed

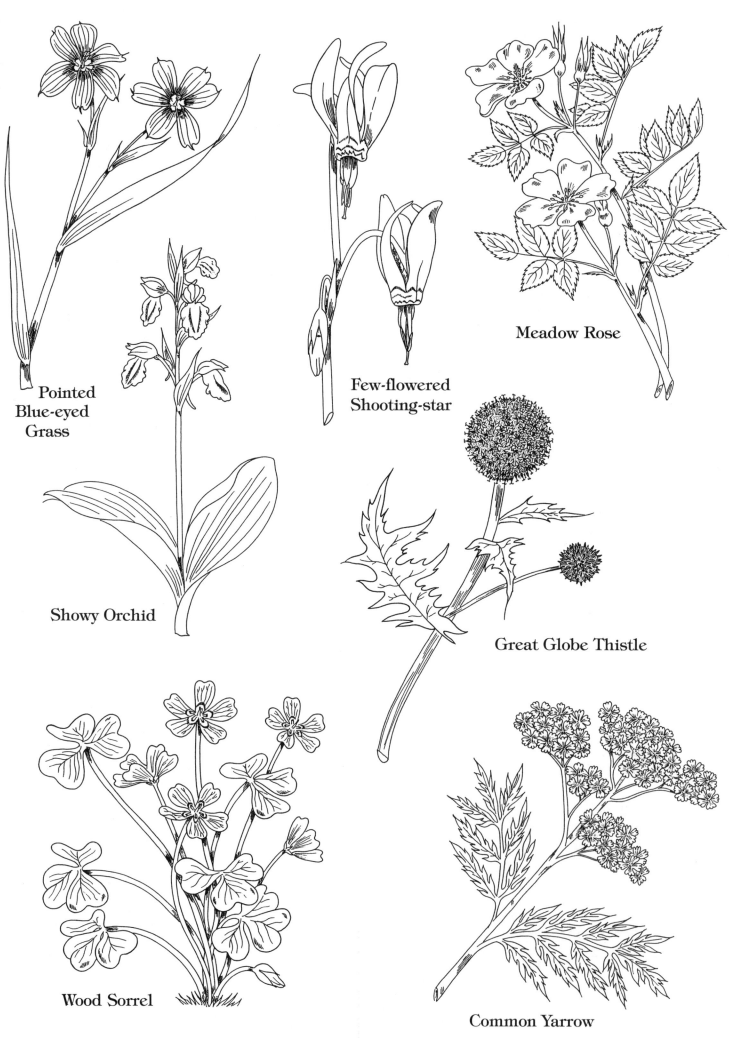

Pointed
Blue-eyed
Grass

Showy Orchid

Few-flowered
Shooting-star

Meadow Rose

Great Globe Thistle

Wood Sorrel

Common Yarrow

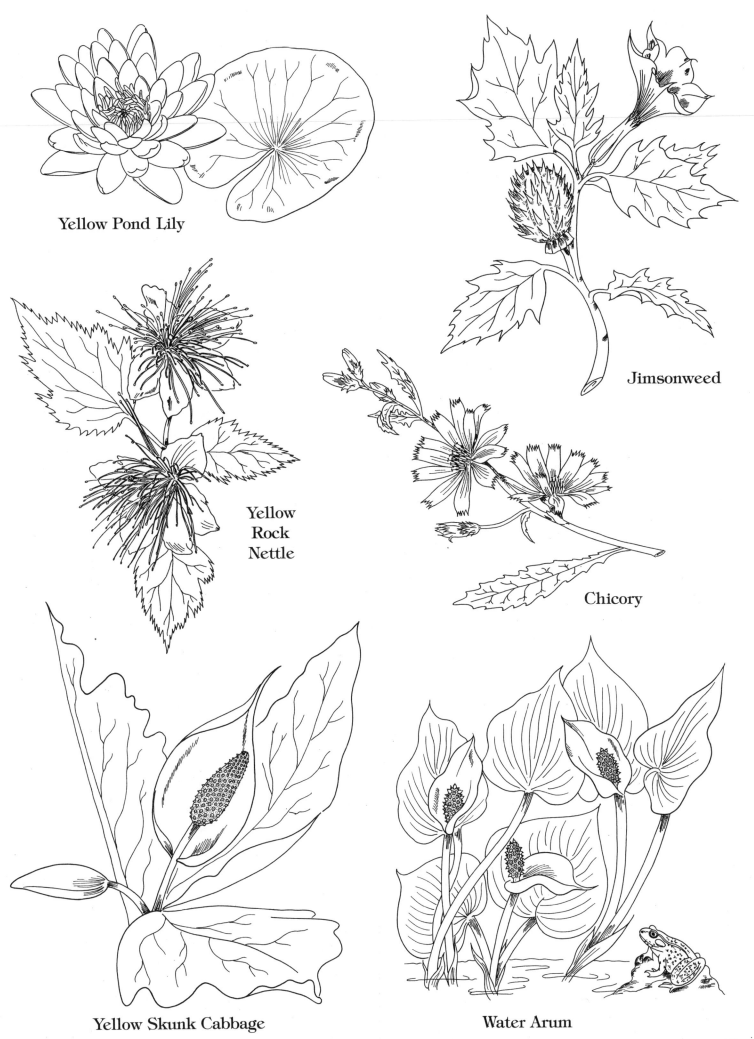

Yellow Pond Lily

Jimsonweed

Yellow
Rock
Nettle

Chicory

Yellow Skunk Cabbage

Water Arum

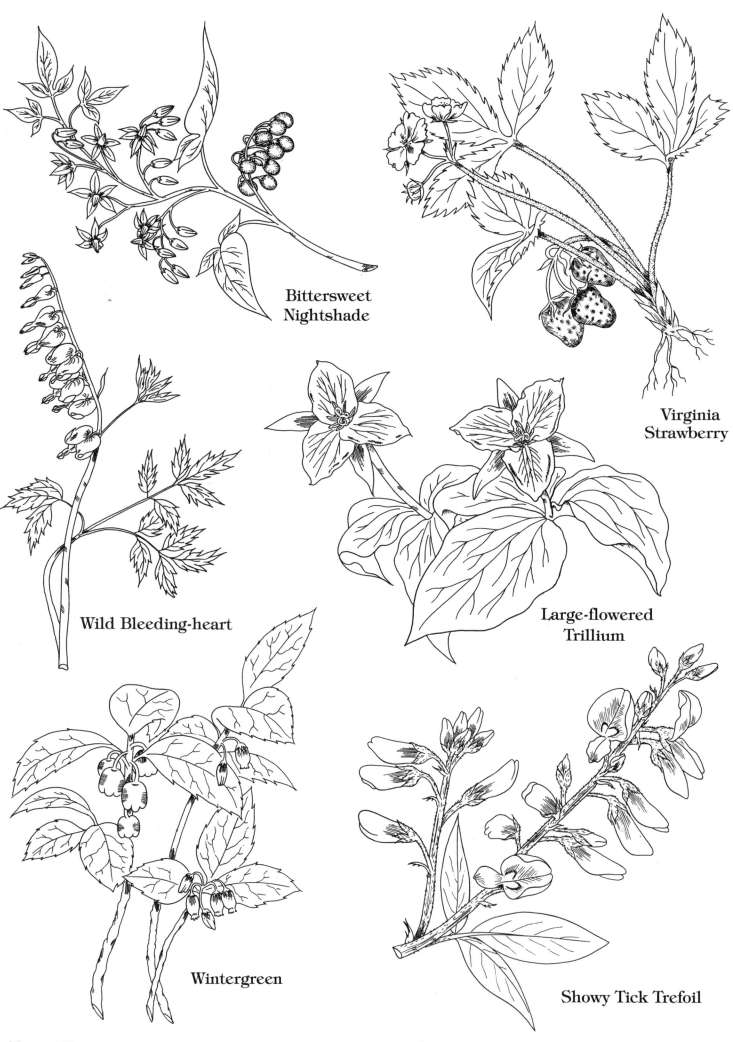

Bittersweet
Nightshade

Virginia
Strawberry

Wild Bleeding-heart

Large-flowered
Trillium

Wintergreen

Showy Tick Trefoil

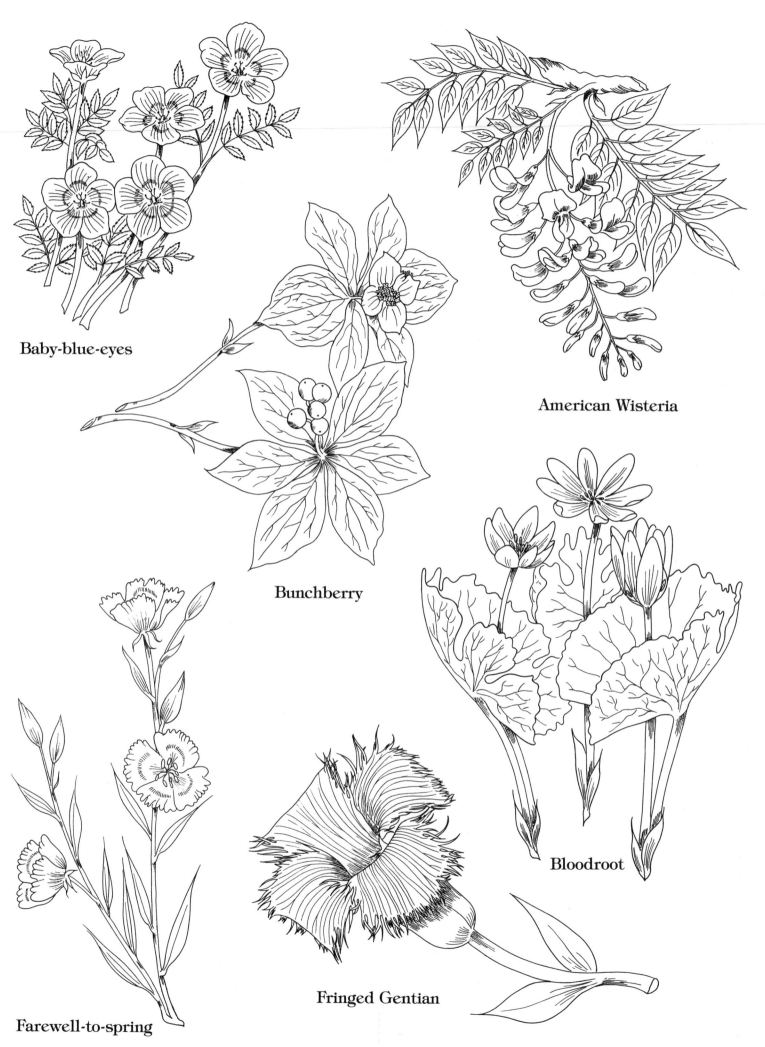

Baby-blue-eyes

Bunchberry

American Wisteria

Bloodroot

Farewell-to-spring

Fringed Gentian

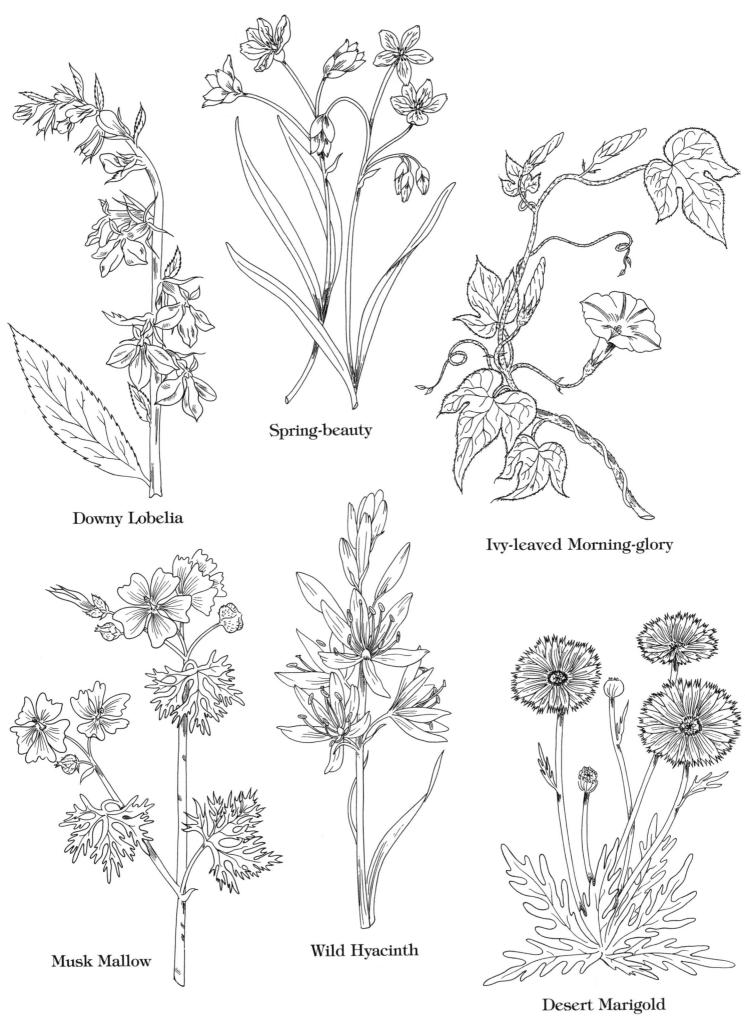

Spring-beauty

Downy Lobelia

Ivy-leaved Morning-glory

Musk Mallow

Wild Hyacinth

Desert Marigold

Mosaic plant

Schefflera

Areca Palm

Baby's-tears

Aglaeonema

Cineraria

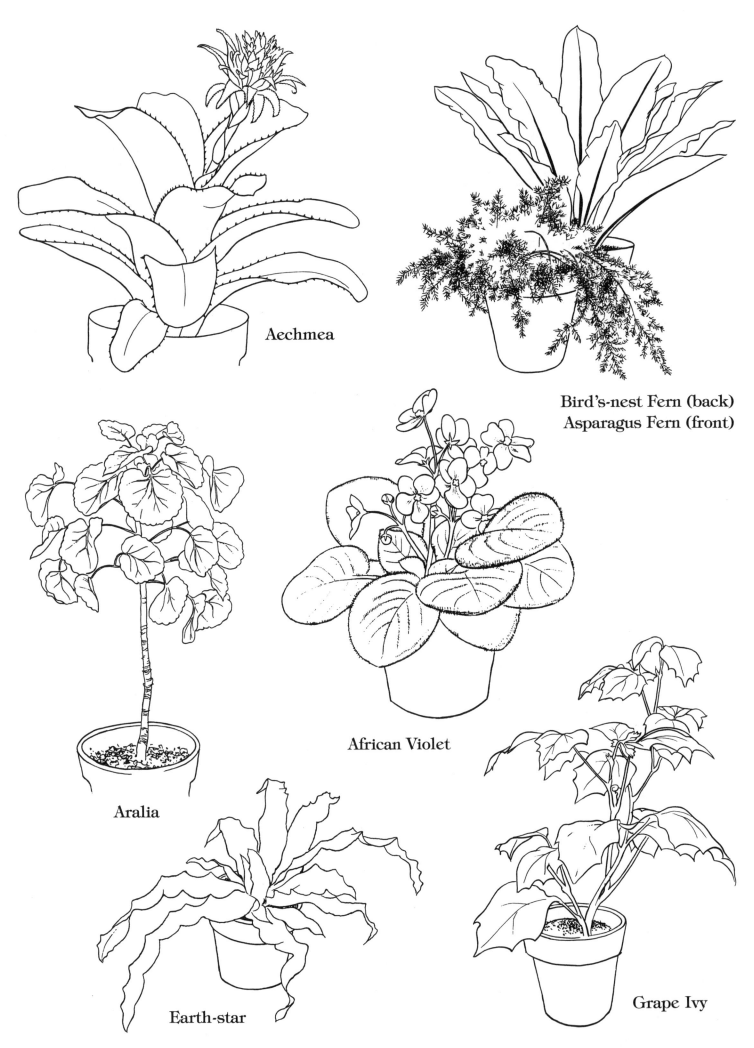

Aechmea

Bird's-nest Fern (back)
Asparagus Fern (front)

Aralia

African Violet

Earth-star

Grape Ivy

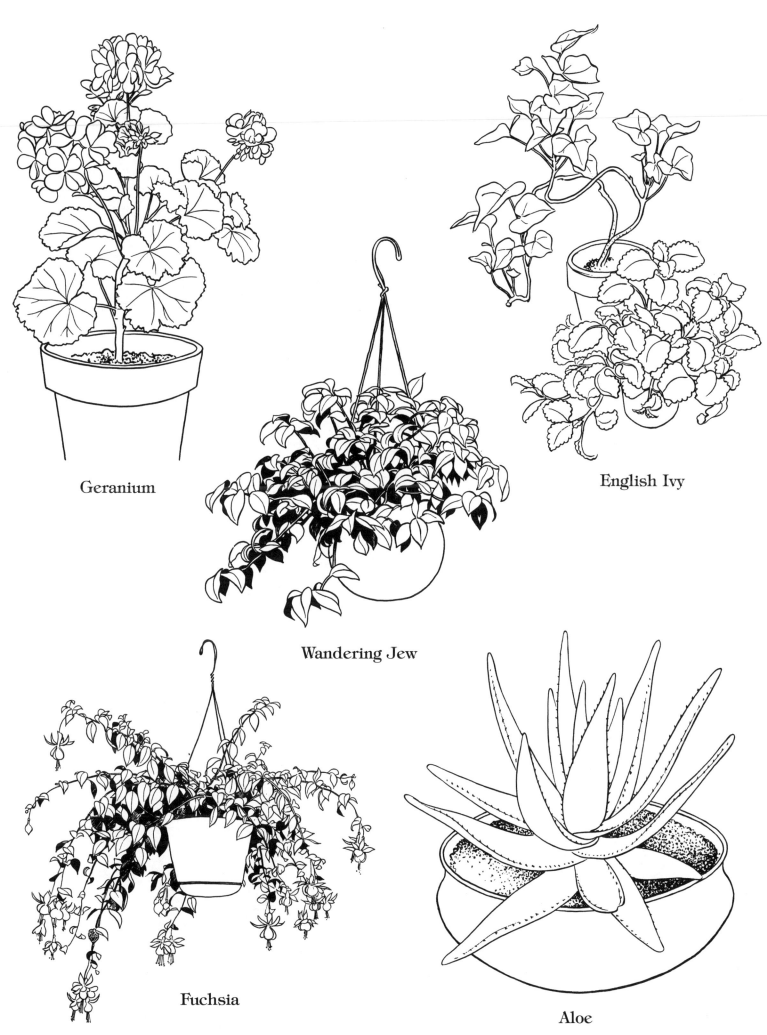

Geranium

Wandering Jew

English Ivy

Fuchsia

Aloe

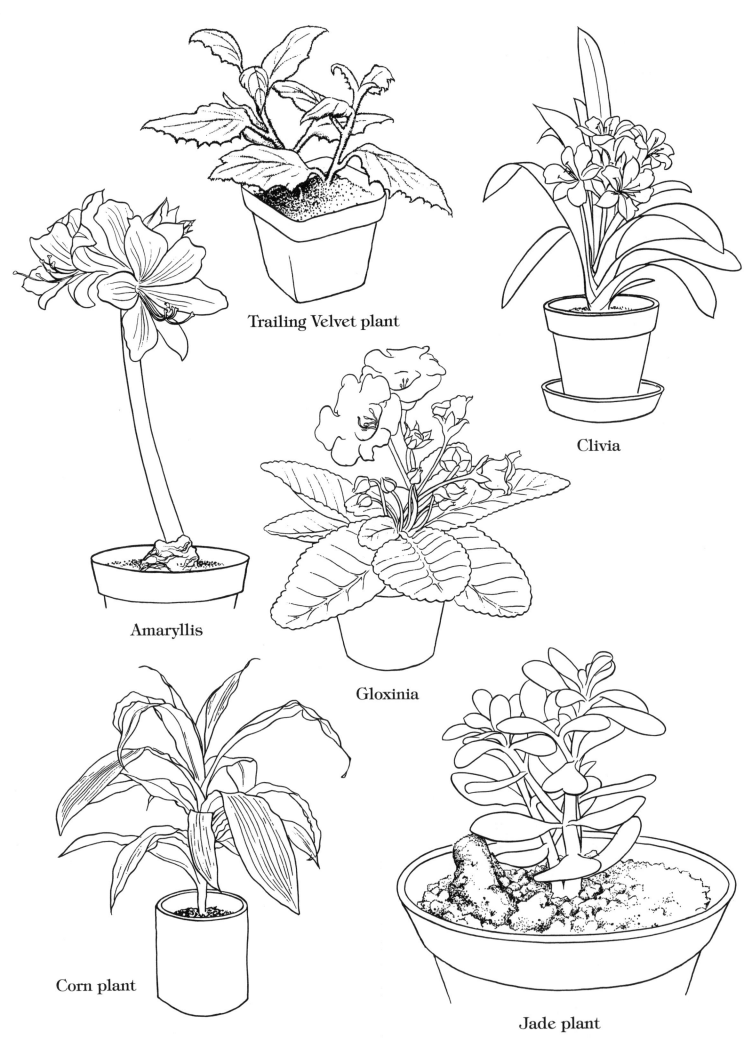

Trailing Velvet plant

Clivia

Amaryllis

Gloxinia

Corn plant

Jade plant

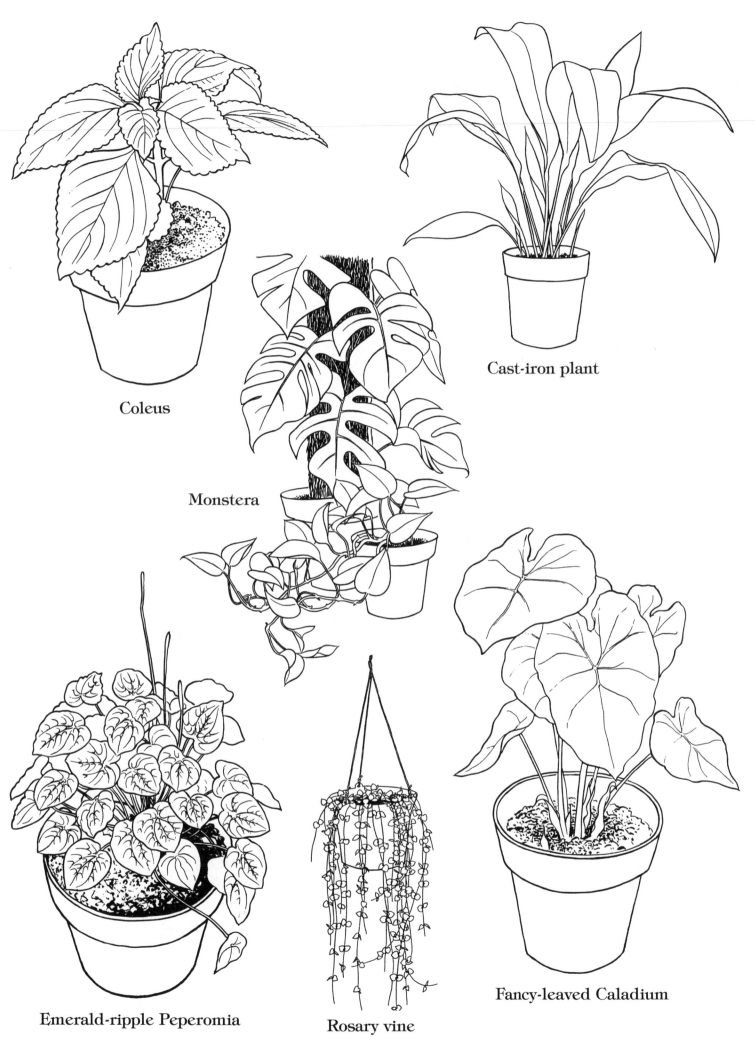

Coleus

Cast-iron plant

Monstera

Emerald-ripple Peperomia

Rosary vine

Fancy-leaved Caladium

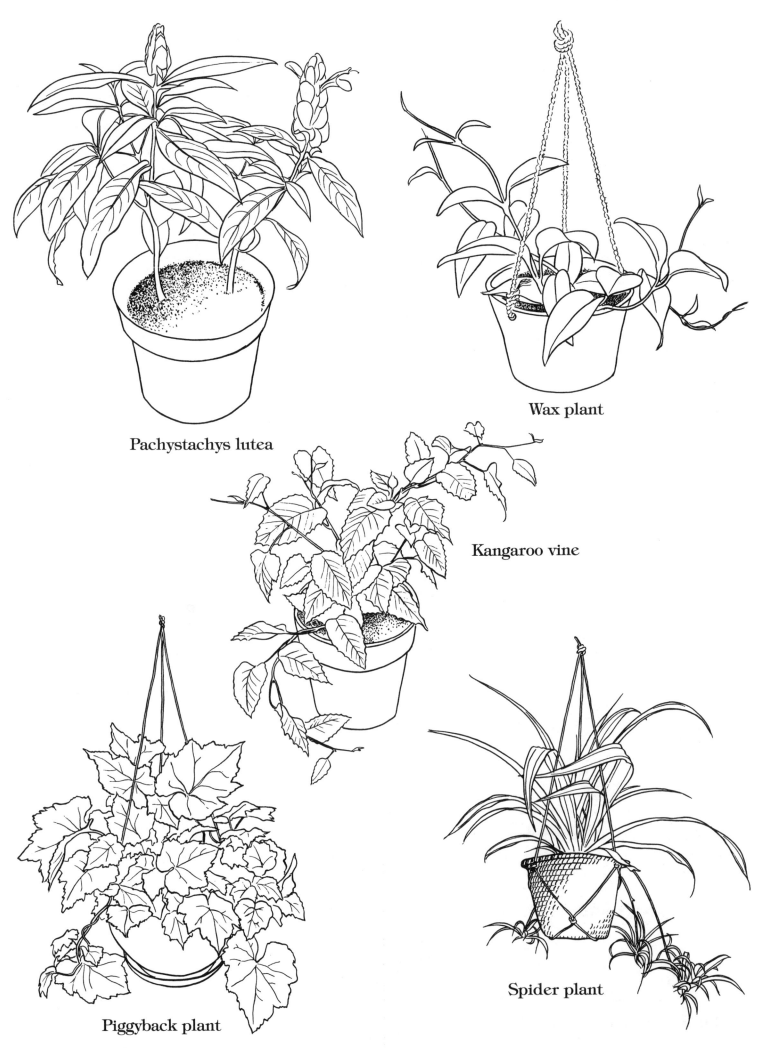

Pachystachys lutea

Wax plant

Kangaroo vine

Piggyback plant

Spider plant

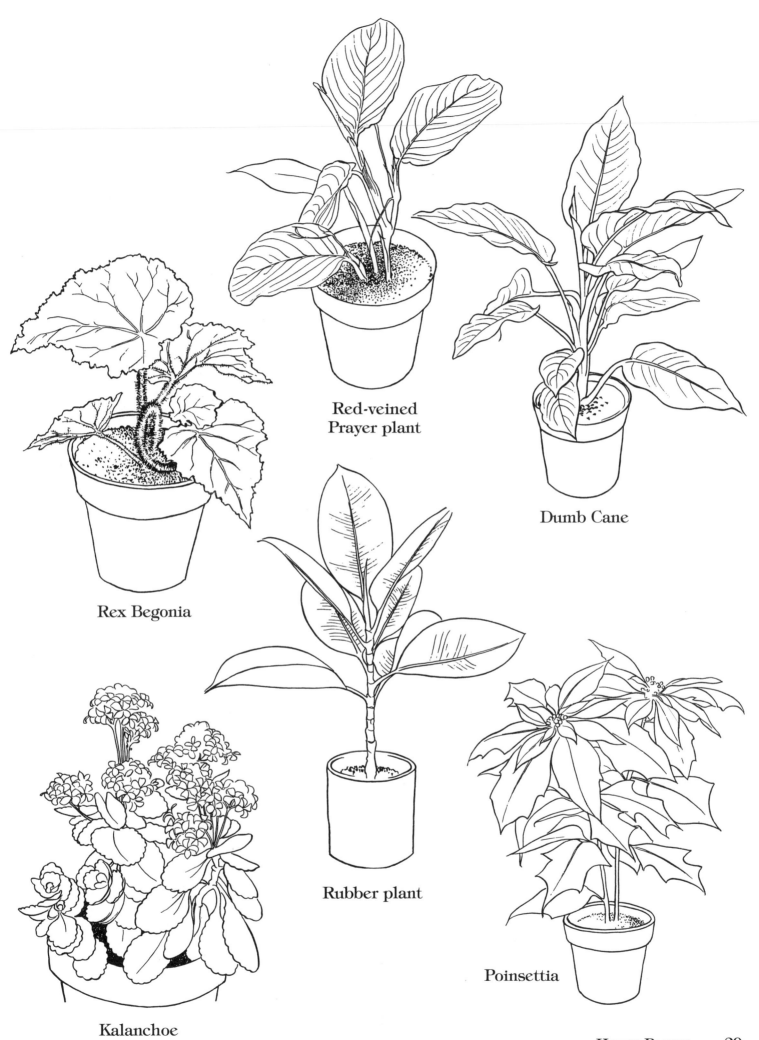

Red-veined
Prayer plant

Dumb Cane

Rex Begonia

Rubber plant

Poinsettia

Kalanchoe

Marjoram

Mallow

Lemon Verbena

Parsley

Lavender

Sweet Bay

Thyme

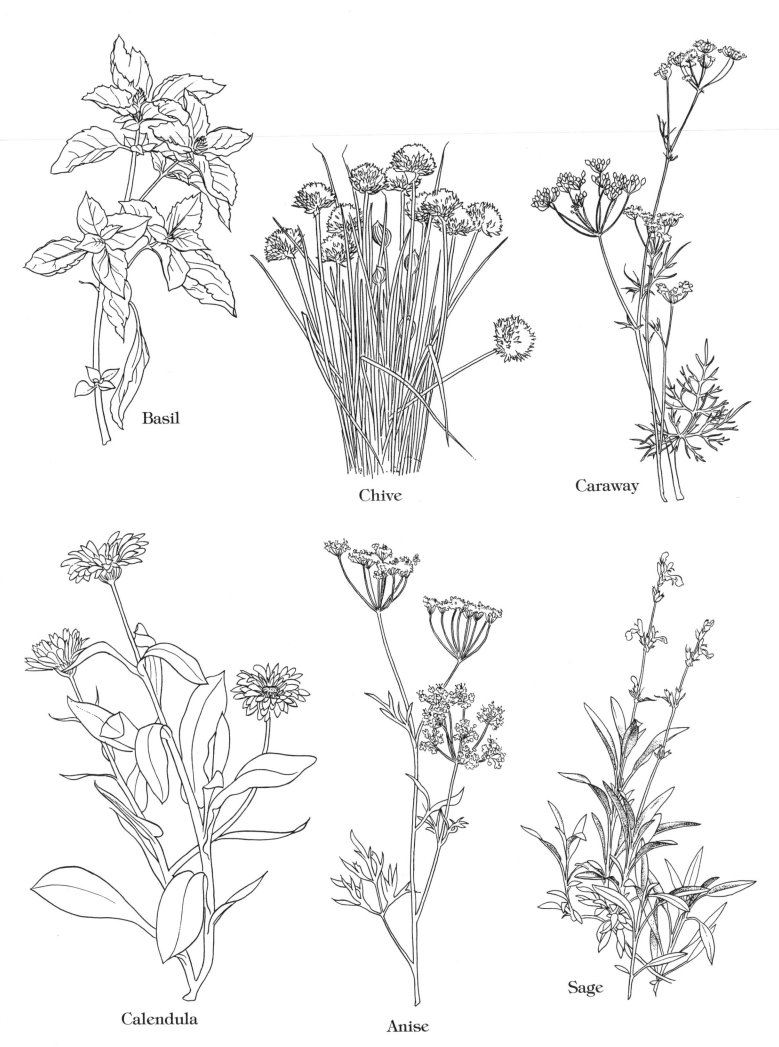

Basil

Chive

Caraway

Calendula

Anise

Sage

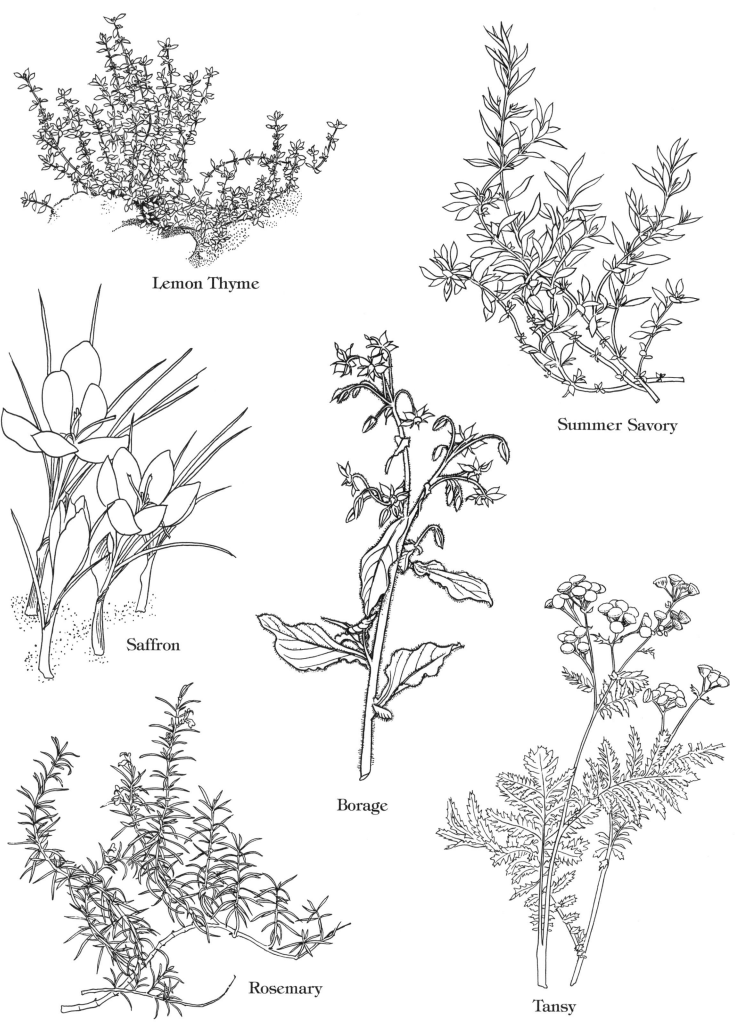

Lemon Thyme

Summer Savory

Saffron

Borage

Rosemary

Tansy

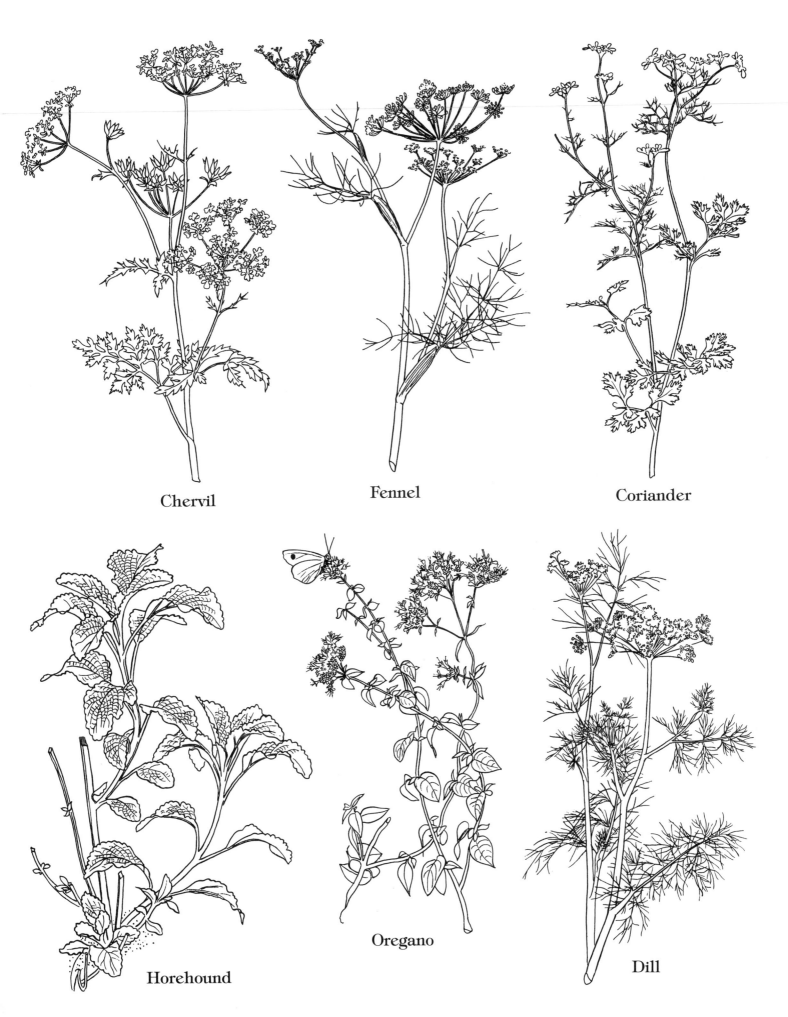

Chervil

Fennel

Coriander

Horehound

Oregano

Dill

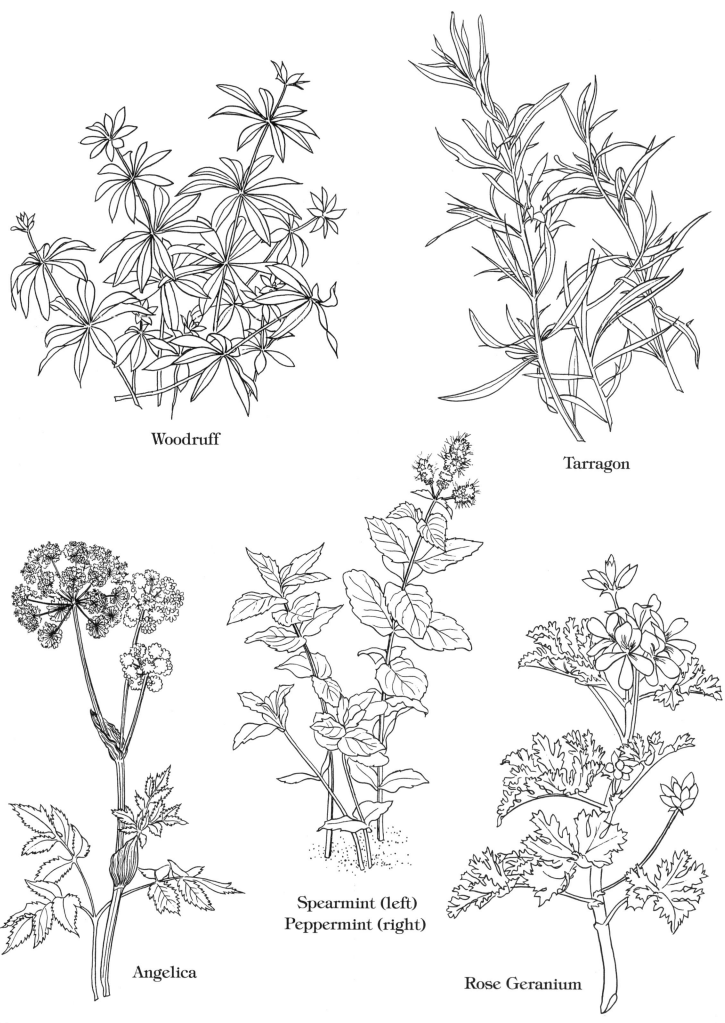

Woodruff

Tarragon

Spearmint (left)
Peppermint (right)

Angelica

Rose Geranium

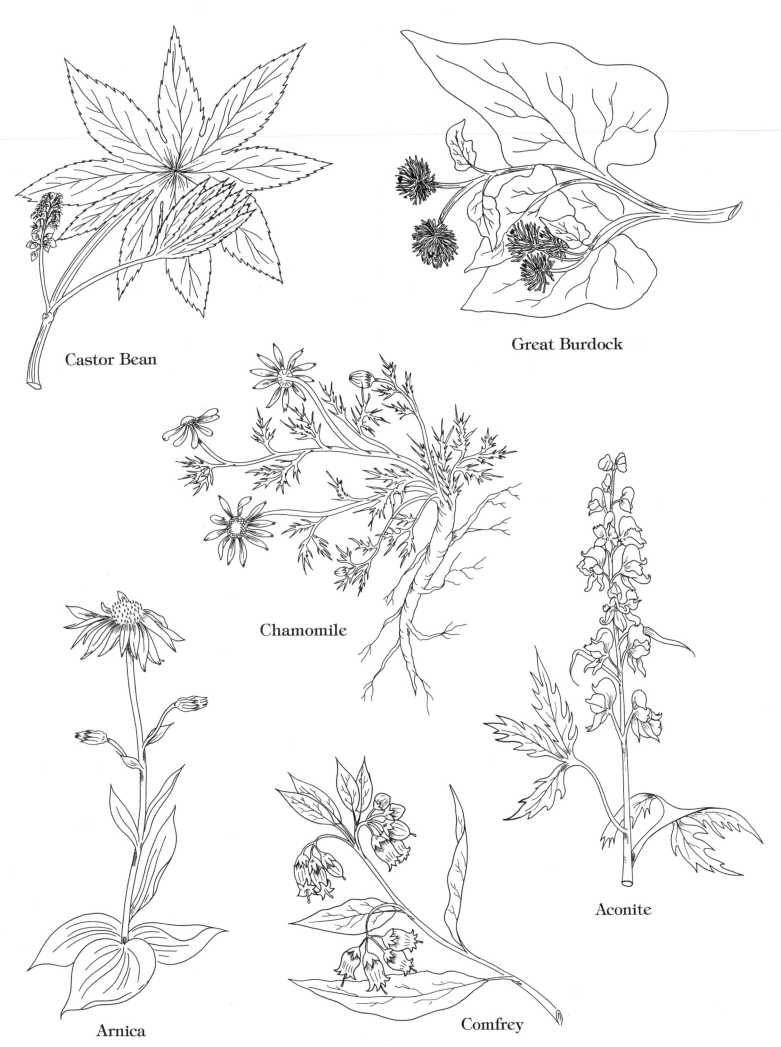

Castor Bean

Great Burdock

Chamomile

Arnica

Comfrey

Aconite

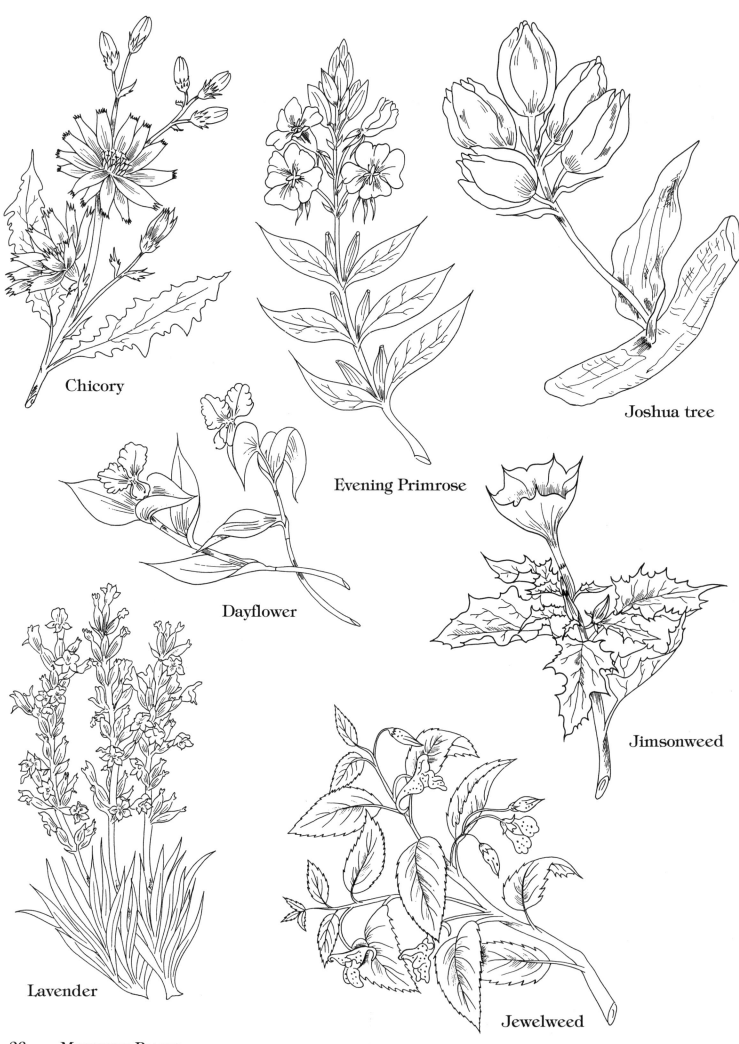

Chicory

Joshua tree

Evening Primrose

Dayflower

Jimsonweed

Lavender

Jewelweed

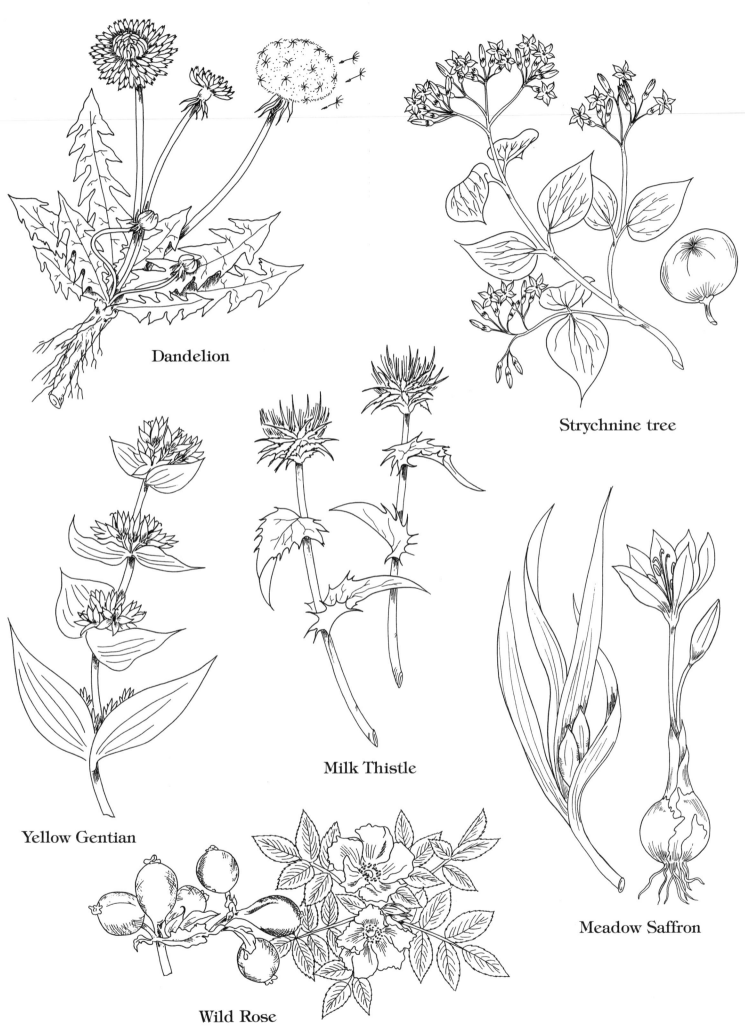

Dandelion

Strychnine tree

Milk Thistle

Yellow Gentian

Meadow Saffron

Wild Rose

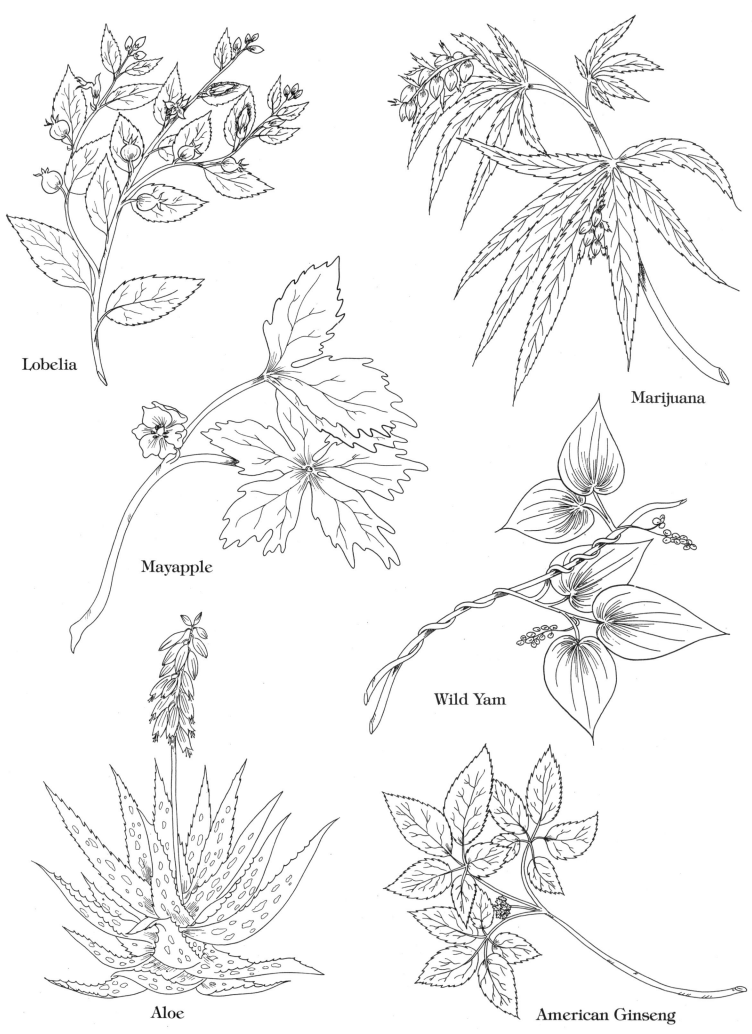

Lobelia

Marijuana

Mayapple

Wild Yam

Aloe

American Ginseng

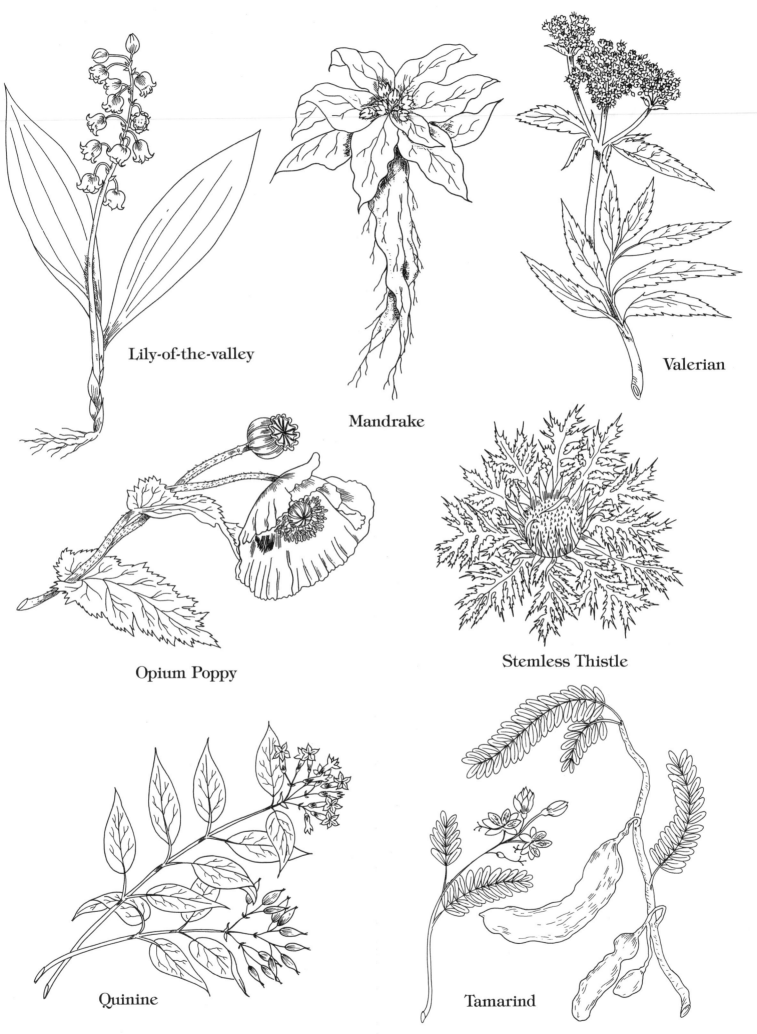

Lily-of-the-valley

Mandrake

Valerian

Opium Poppy

Stemless Thistle

Quinine

Tamarind

Pomegranate

Squill

Elecampane

Foxglove

Passionflower

Marigold

Balm-of-Gilead

Hedge Bindweed

Coca

Boneset

Cornflower

Belladonna

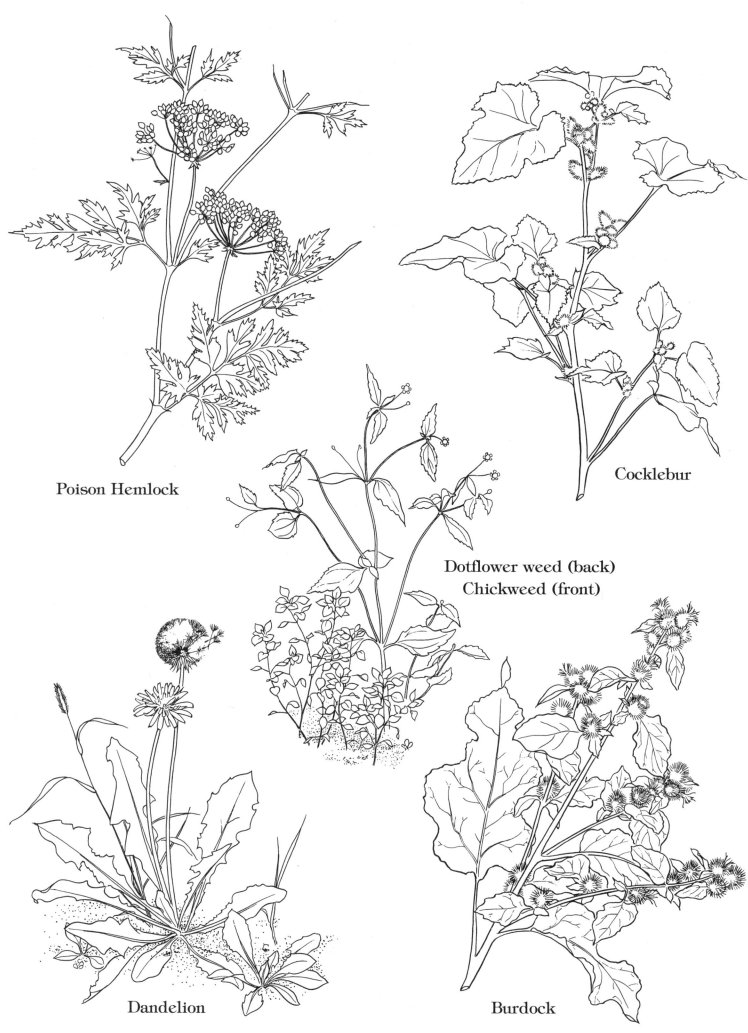

Poison Hemlock

Cocklebur

Dotflower weed (back)
Chickweed (front)

Dandelion

Burdock

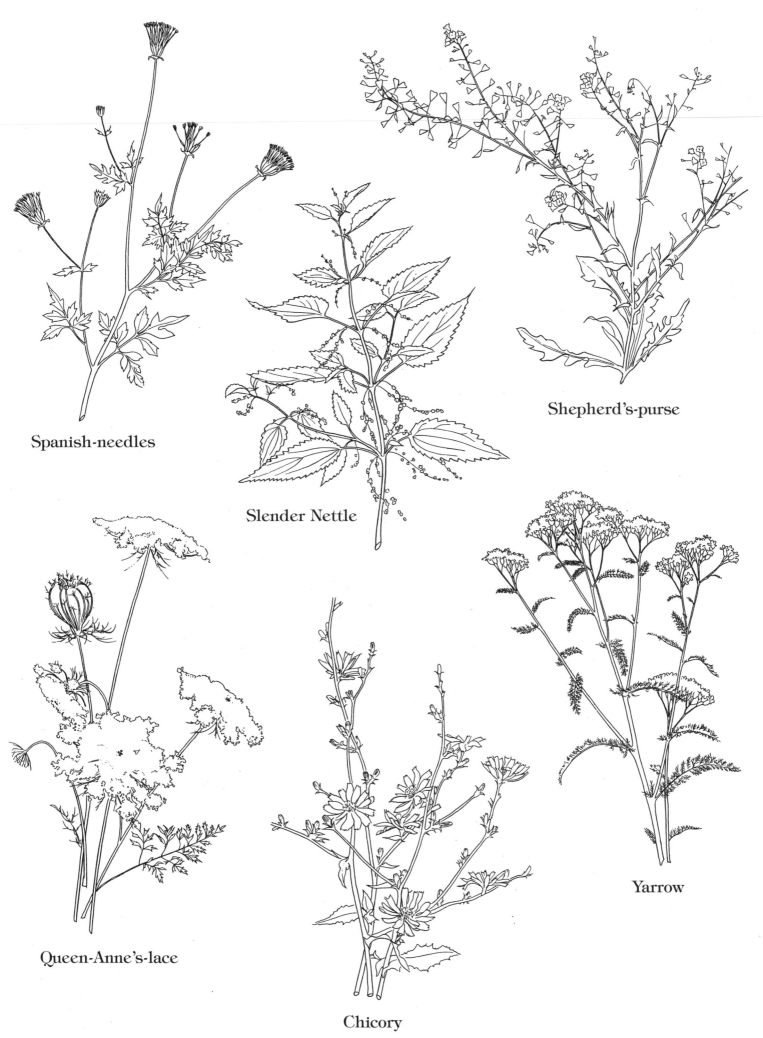

Spanish-needles

Slender Nettle

Shepherd's-purse

Queen-Anne's-lace

Chicory

Yarrow

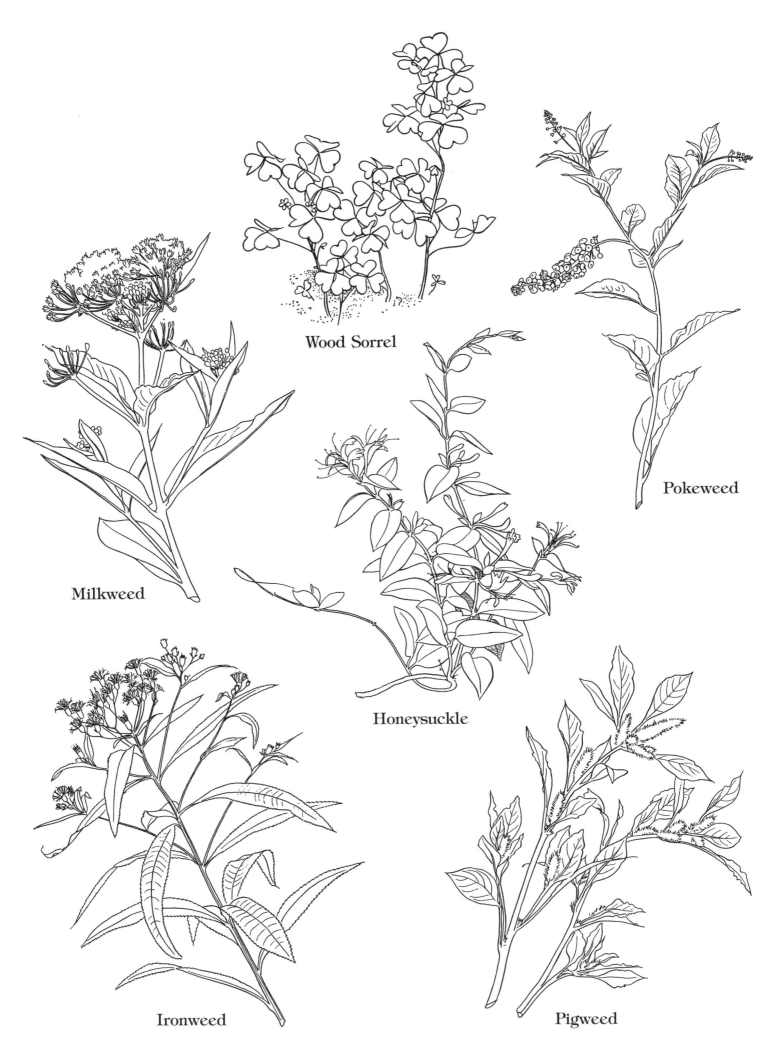

Wood Sorrel

Pokeweed

Milkweed

Honeysuckle

Ironweed

Pigweed

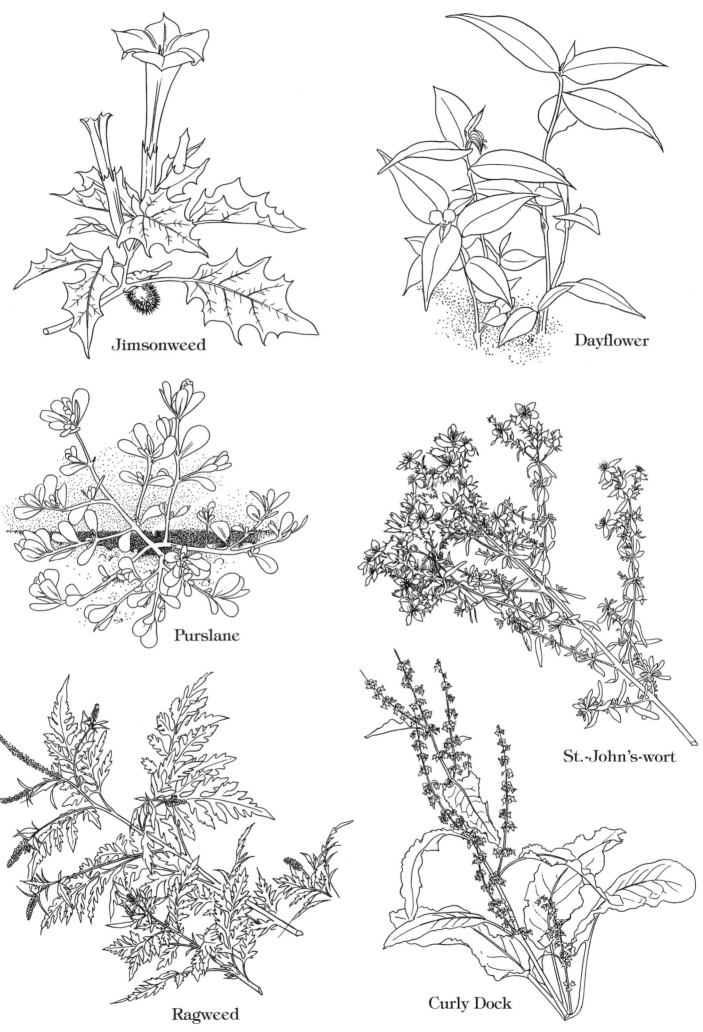

Jimsonweed

Dayflower

Purslane

St.-John's-wort

Ragweed

Curly Dock

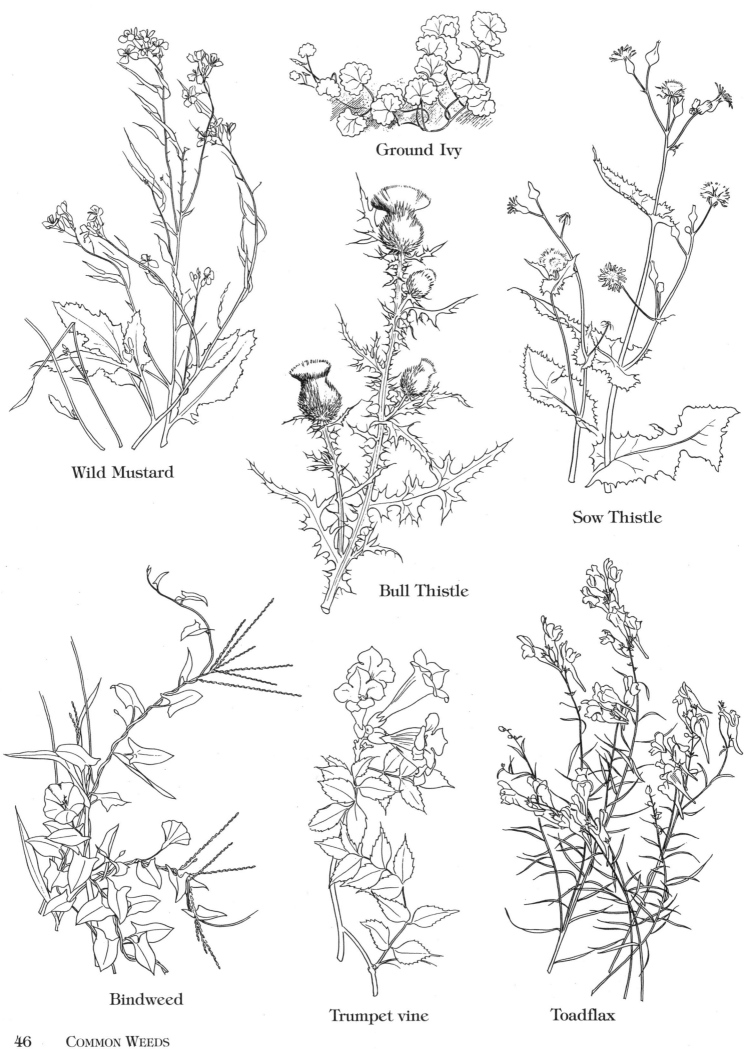

Ground Ivy

Wild Mustard

Bull Thistle

Sow Thistle

Bindweed

Trumpet vine

Toadflax

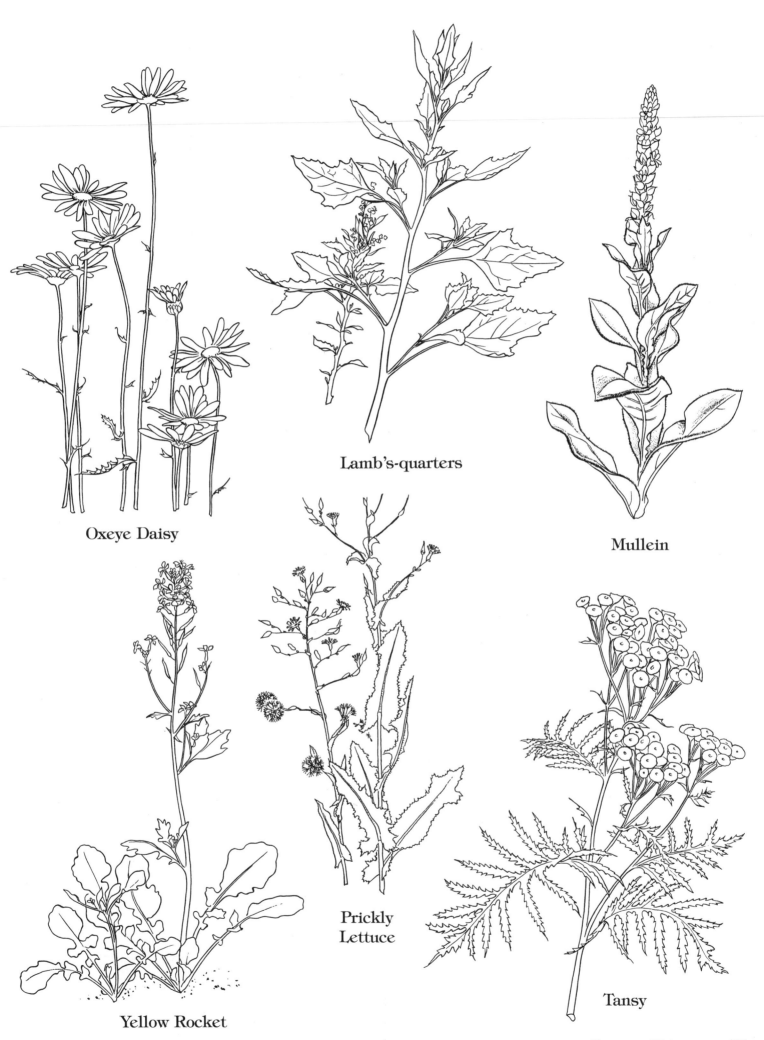

Oxeye Daisy

Lamb's-quarters

Mullein

Yellow Rocket

Prickly
Lettuce

Tansy

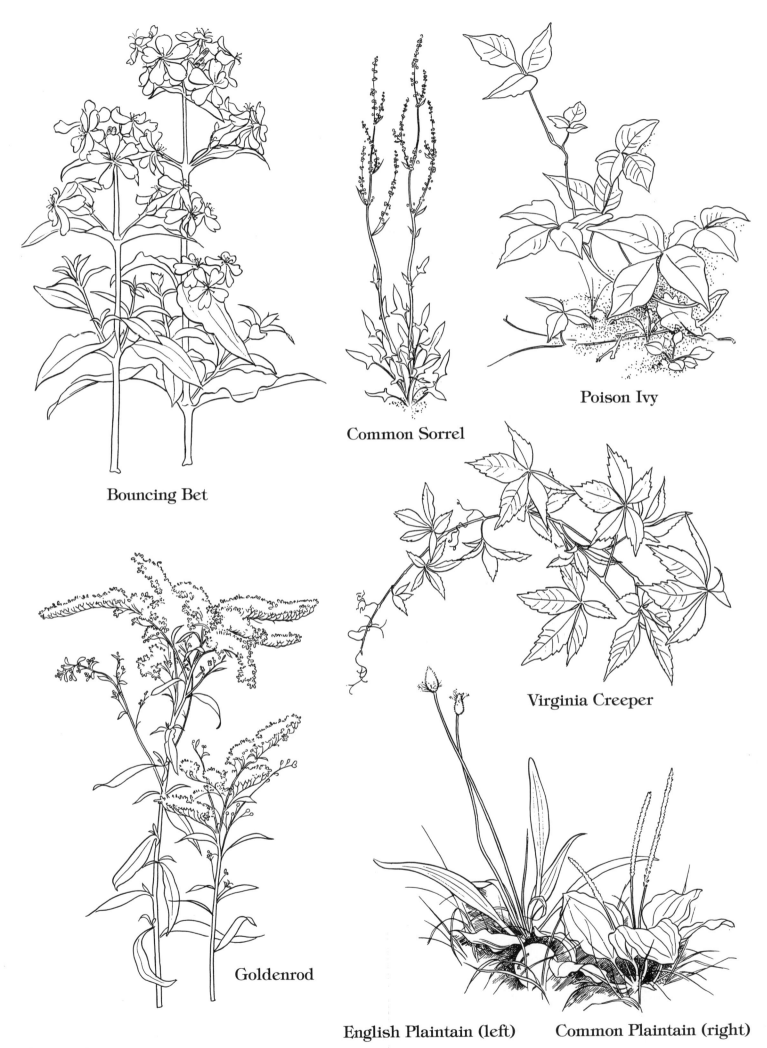

Bouncing Bet

Common Sorrel

Poison Ivy

Virginia Creeper

Goldenrod

English Plaintain (left)   Common Plaintain (right)

Old-lady
Cactus

Brain Cactus

Teddy-bear Cholla

Button Cactus

Rainbow Chin Cactus

Beehive Cactus

Pelecyphora aselliformis

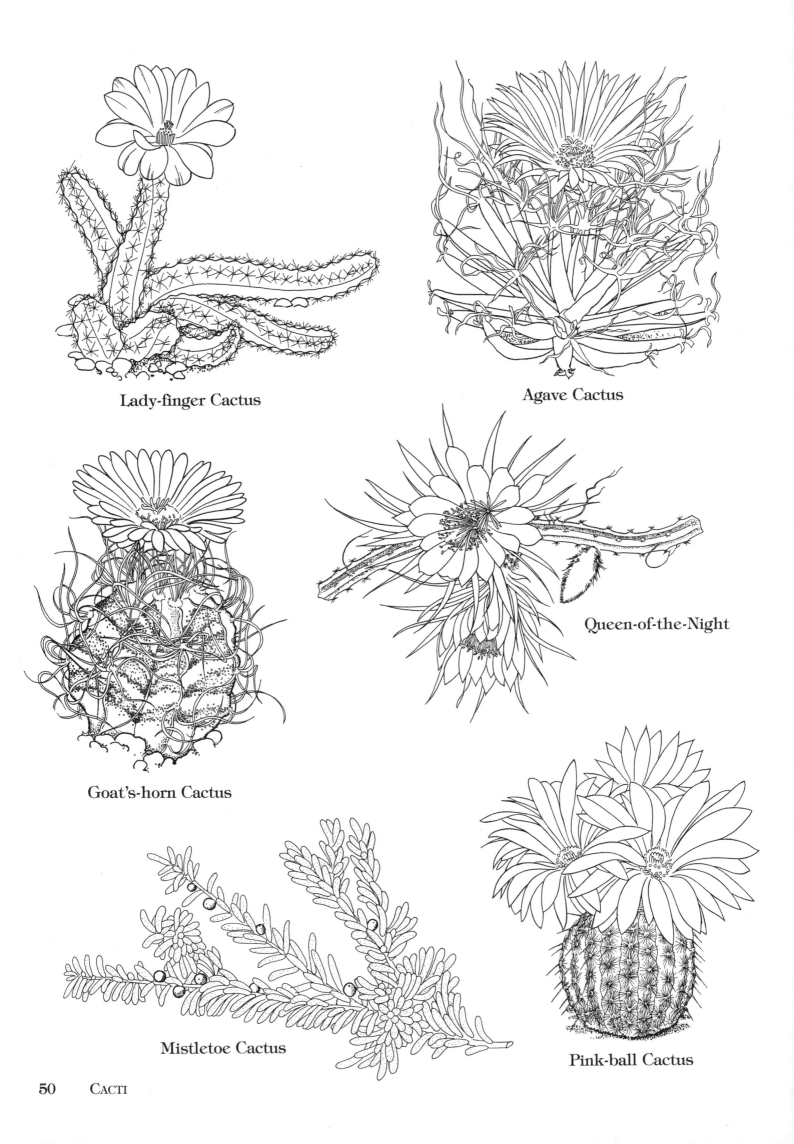

Lady-finger Cactus

Agave Cactus

Goat's-horn Cactus

Queen-of-the-Night

Mistletoe Cactus

Pink-ball Cactus

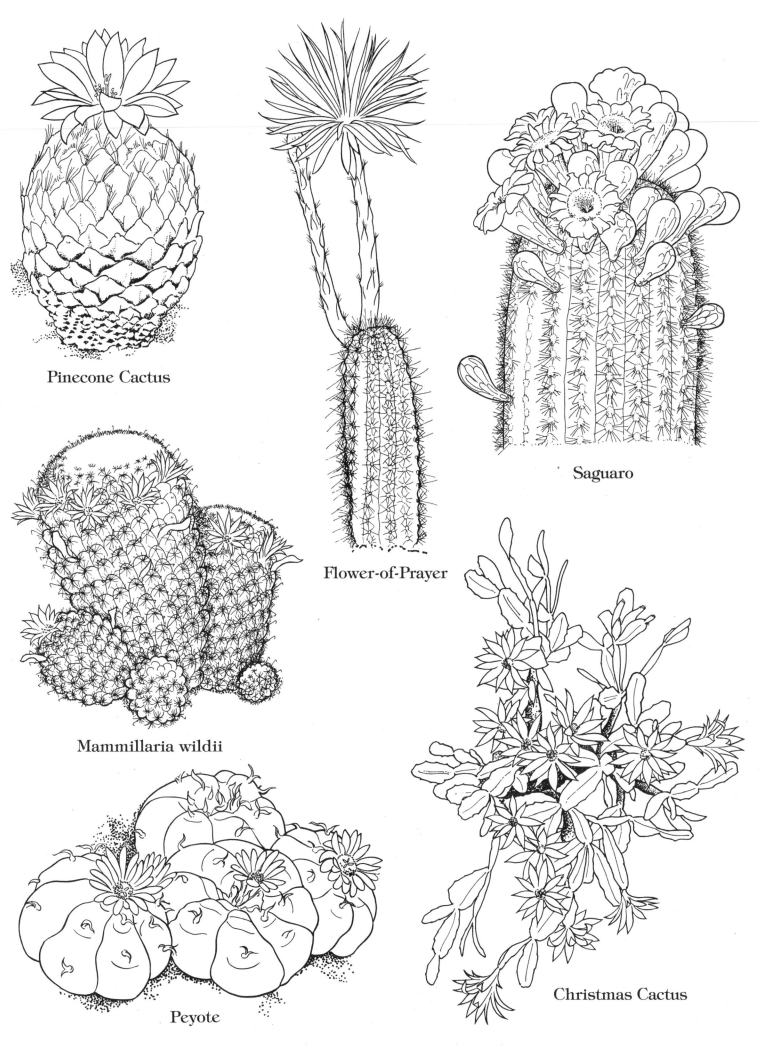

Pinecone Cactus

Flower-of-Prayer

Saguaro

Mammillaria wildii

Peyote

Christmas Cactus

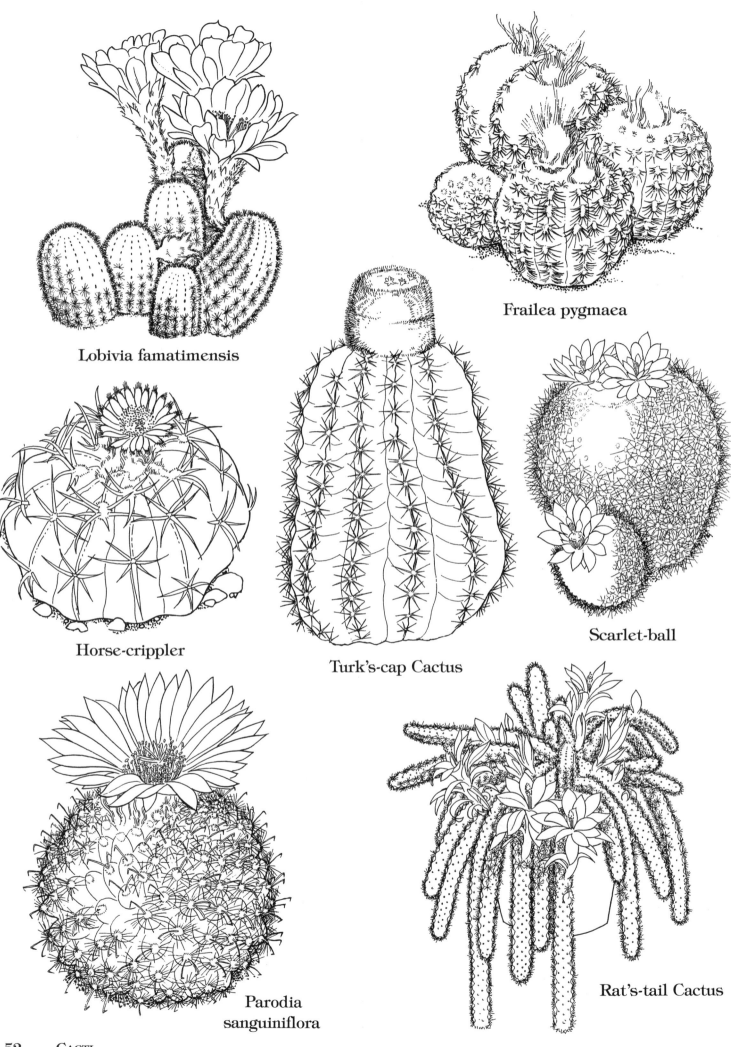

Lobivia famatimensis

Frailea pygmaea

Horse-crippler

Turk's-cap Cactus

Scarlet-ball

Parodia
sanguiniflora

Rat's-tail Cactus

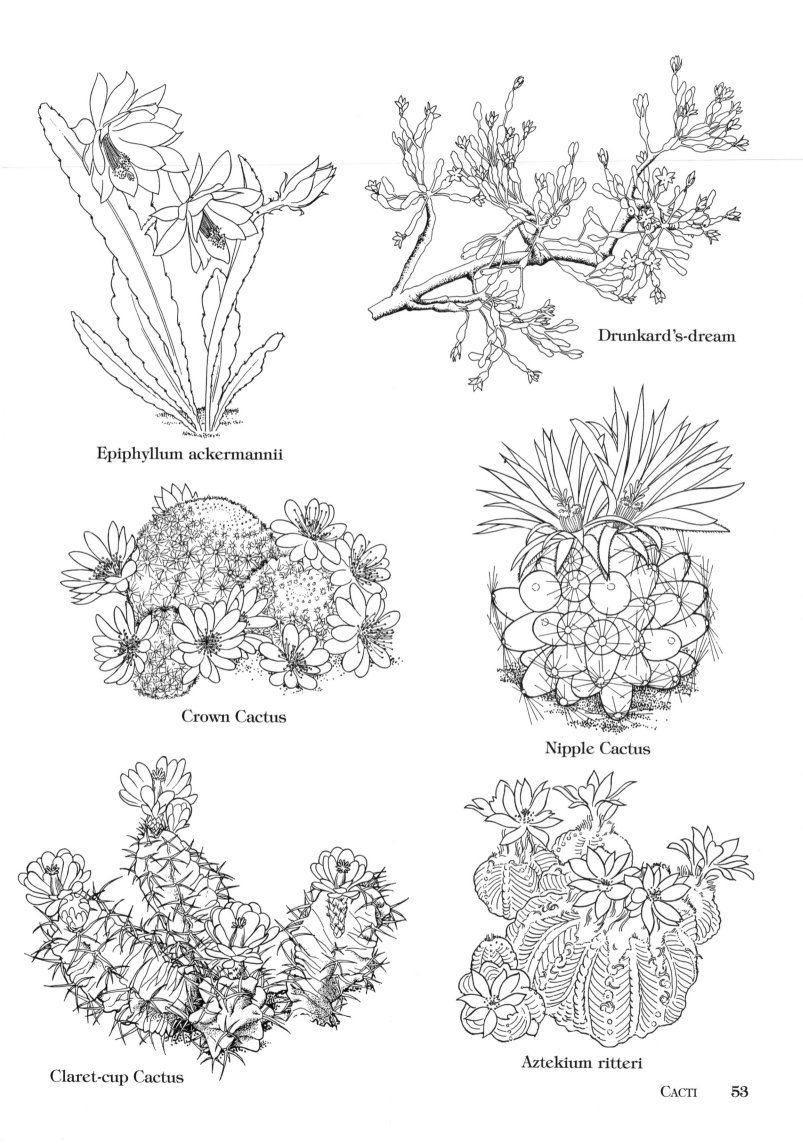

Drunkard's-dream

Epiphyllum ackermannii

Crown Cactus

Nipple Cactus

Claret-cup Cactus

Aztekium ritteri

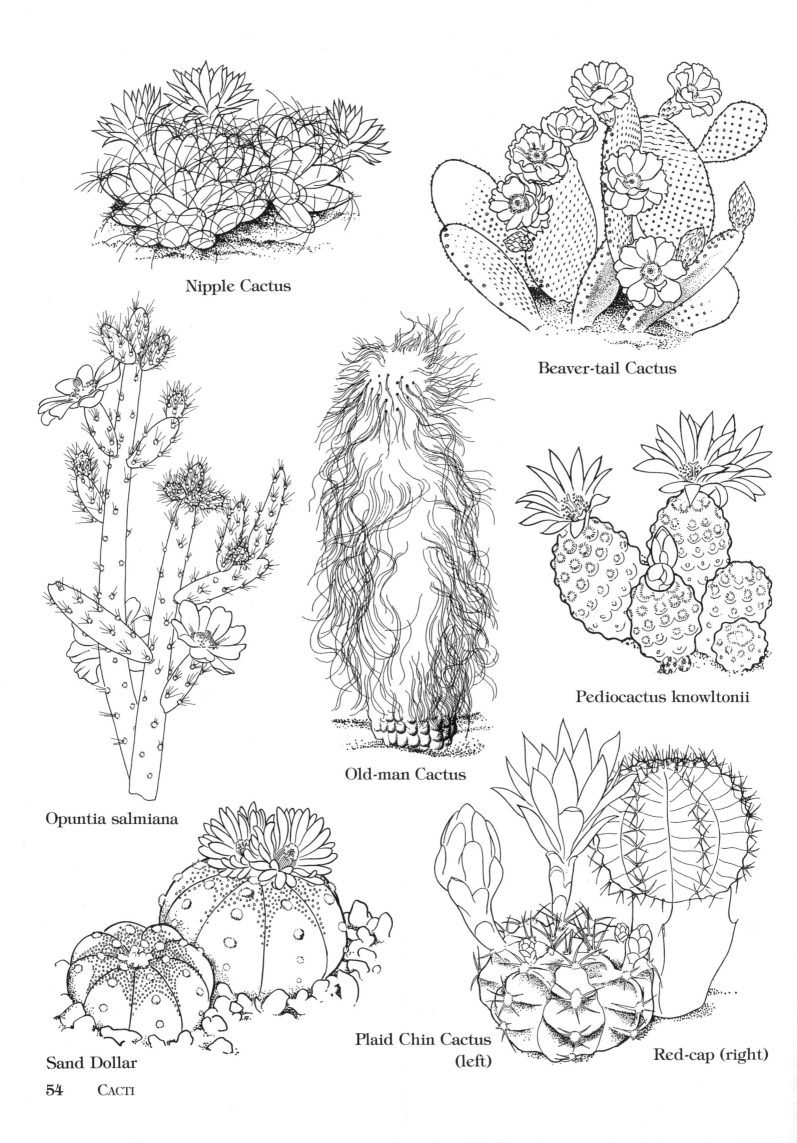

Nipple Cactus

Beaver-tail Cactus

Opuntia salmiana

Old-man Cactus

Pediocactus knowltonii

Sand Dollar

Plaid Chin Cactus
(left)

Red-cap (right)

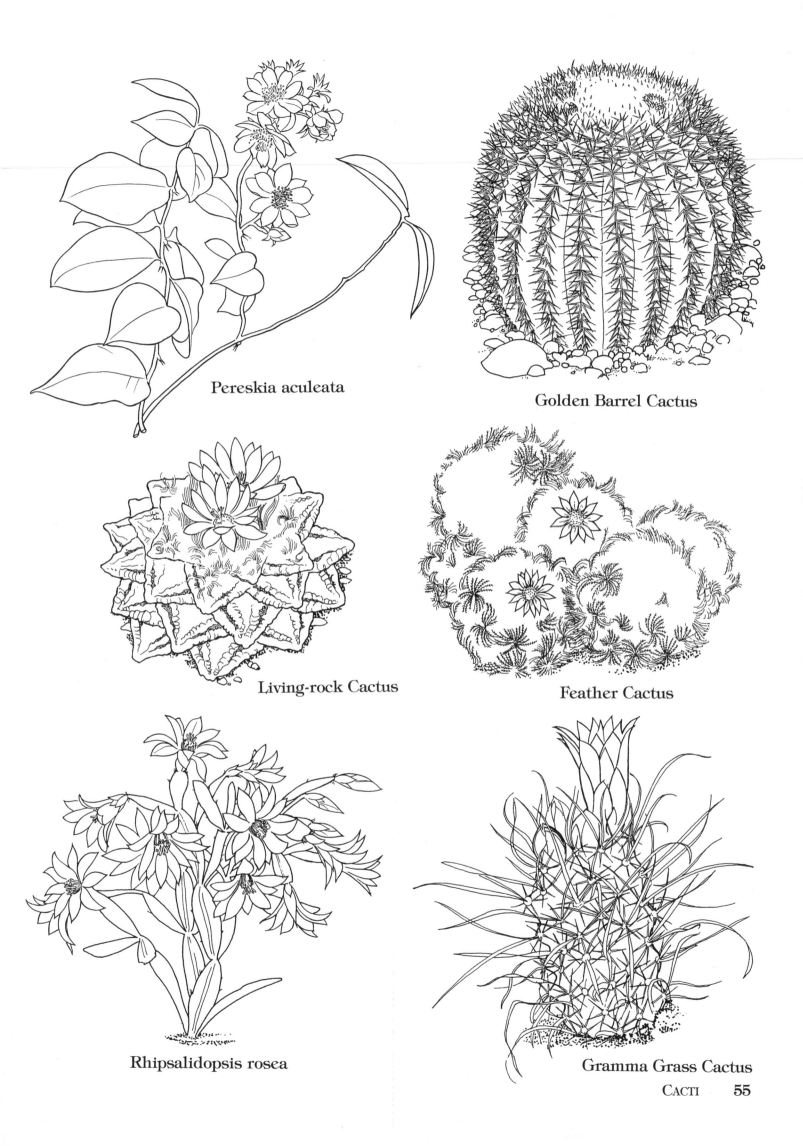

Pereskia aculeata

Golden Barrel Cactus

Living-rock Cactus

Feather Cactus

Rhipsalidopsis rosea

Gramma Grass Cactus

Cat's-claw vine

Tabebuia

Vanda Orchid

Wood Rose

African Daisy

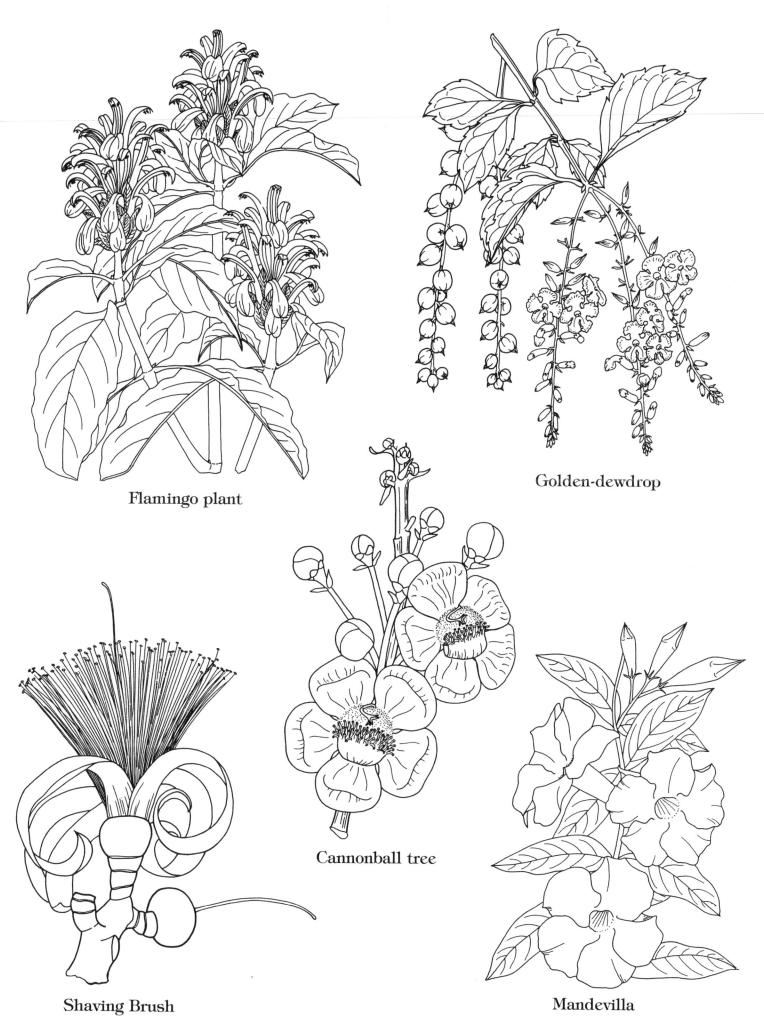

Flamingo plant

Golden-dewdrop

Cannonball tree

Shaving Brush

Mandevilla

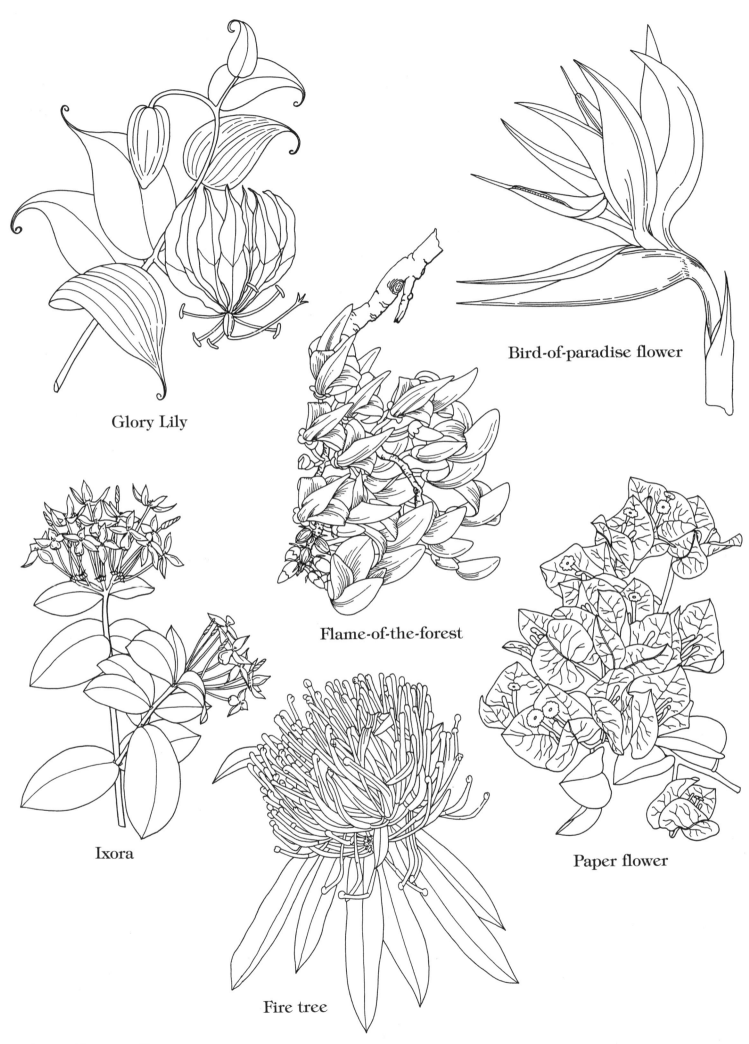

Glory Lily

Bird-of-paradise flower

Flame-of-the-forest

Ixora

Fire tree

Paper flower

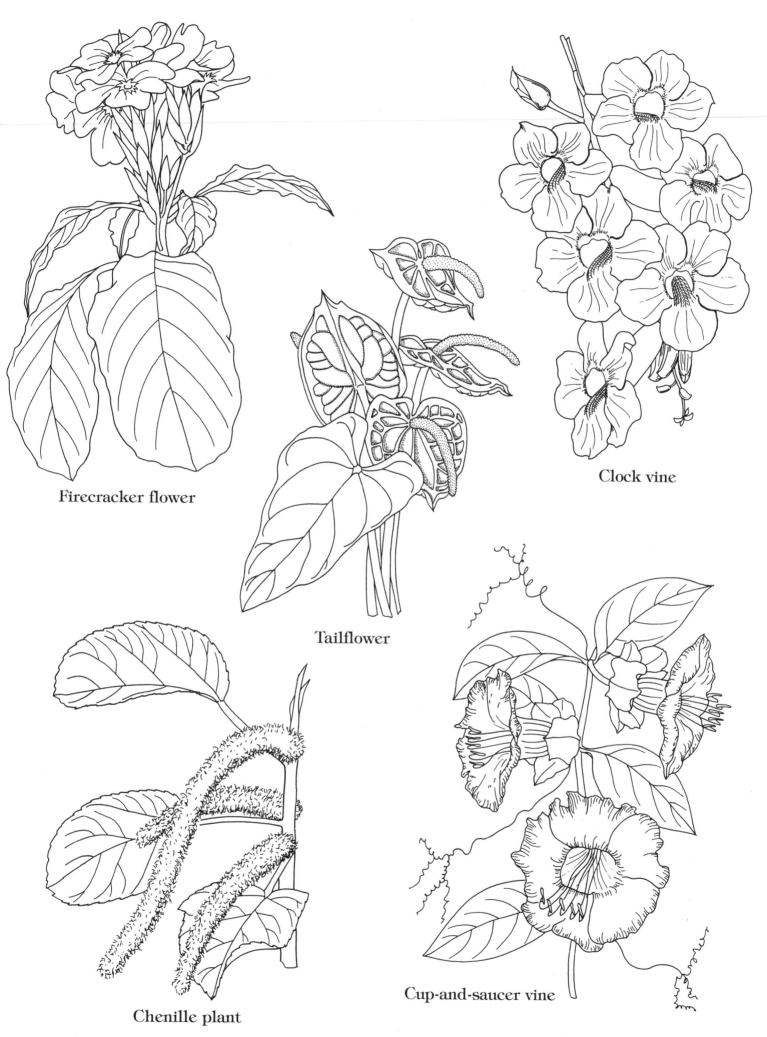

Firecracker flower

Clock vine

Tailflower

Chenille plant

Cup-and-saucer vine

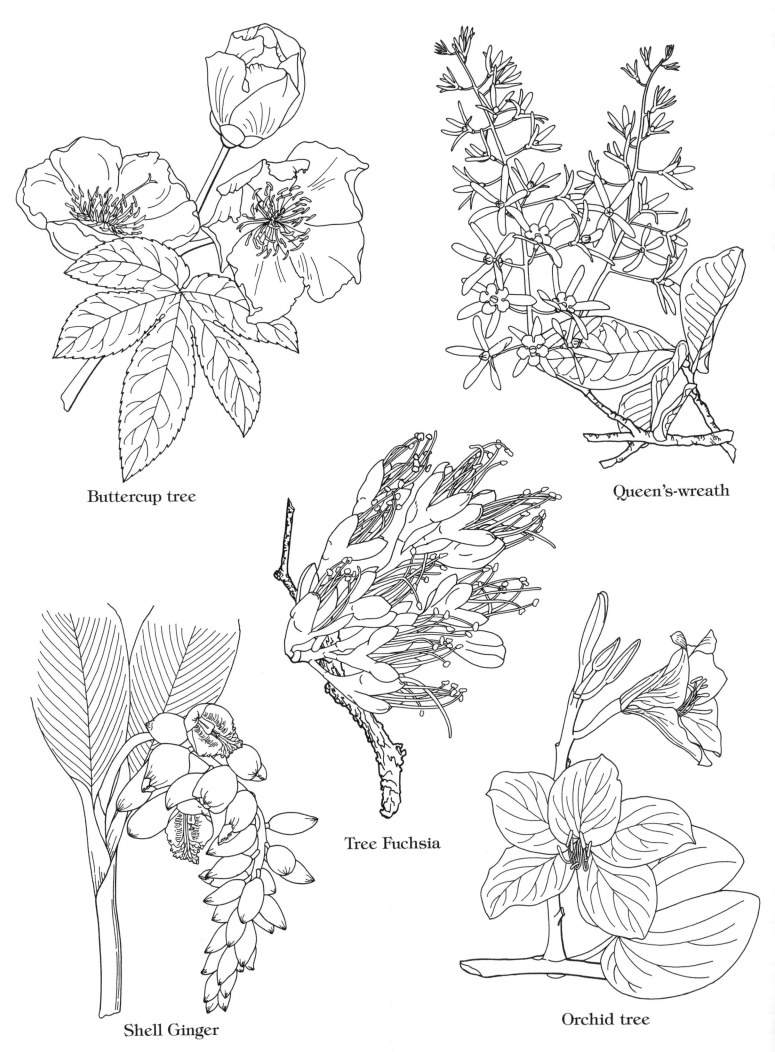

Buttercup tree

Queen's-wreath

Tree Fuchsia

Shell Ginger

Orchid tree

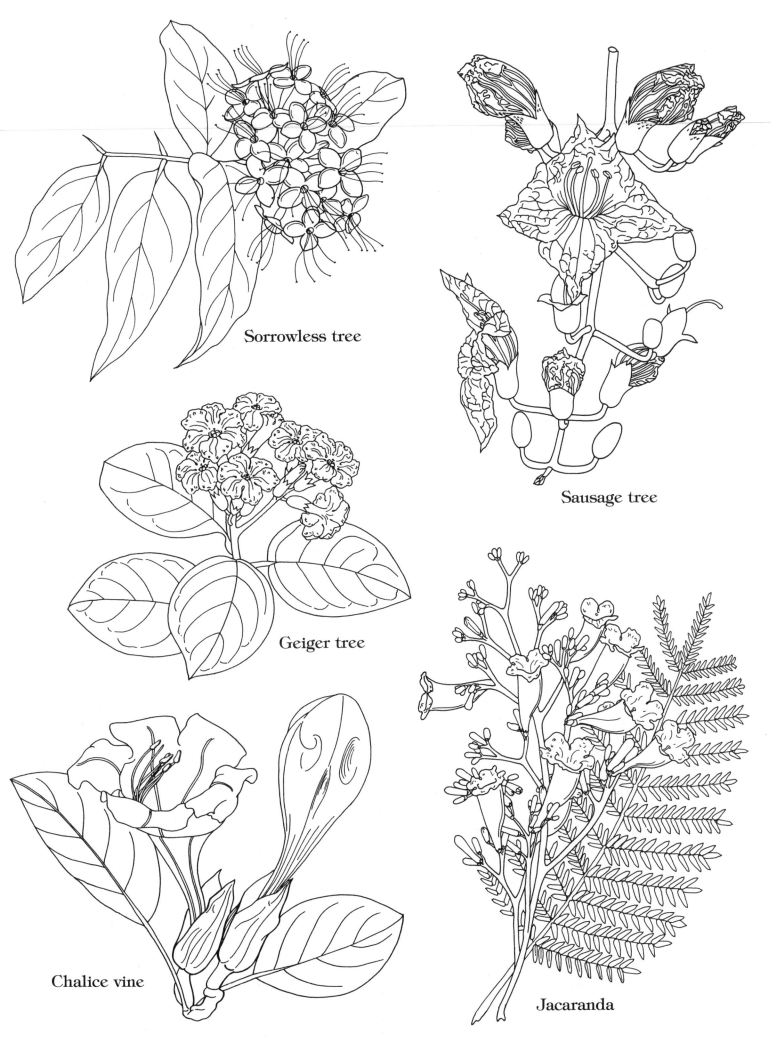

Sorrowless tree

Sausage tree

Geiger tree

Chalice vine

Jacaranda

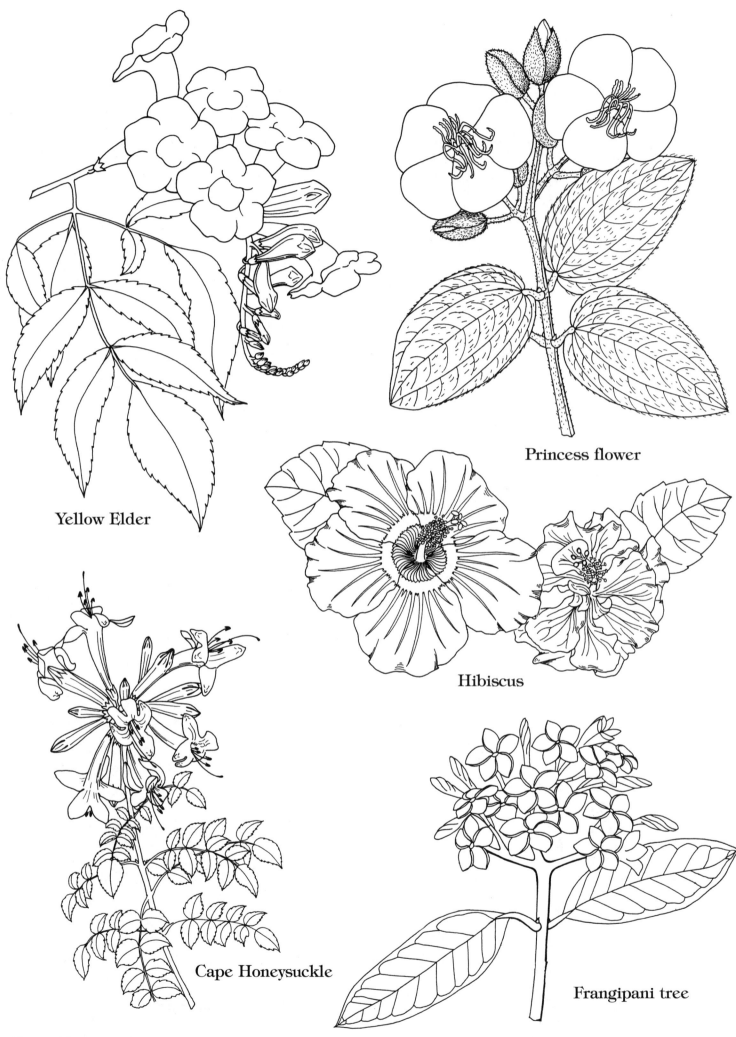

Yellow Elder

Princess flower

Hibiscus

Cape Honeysuckle

Frangipani tree

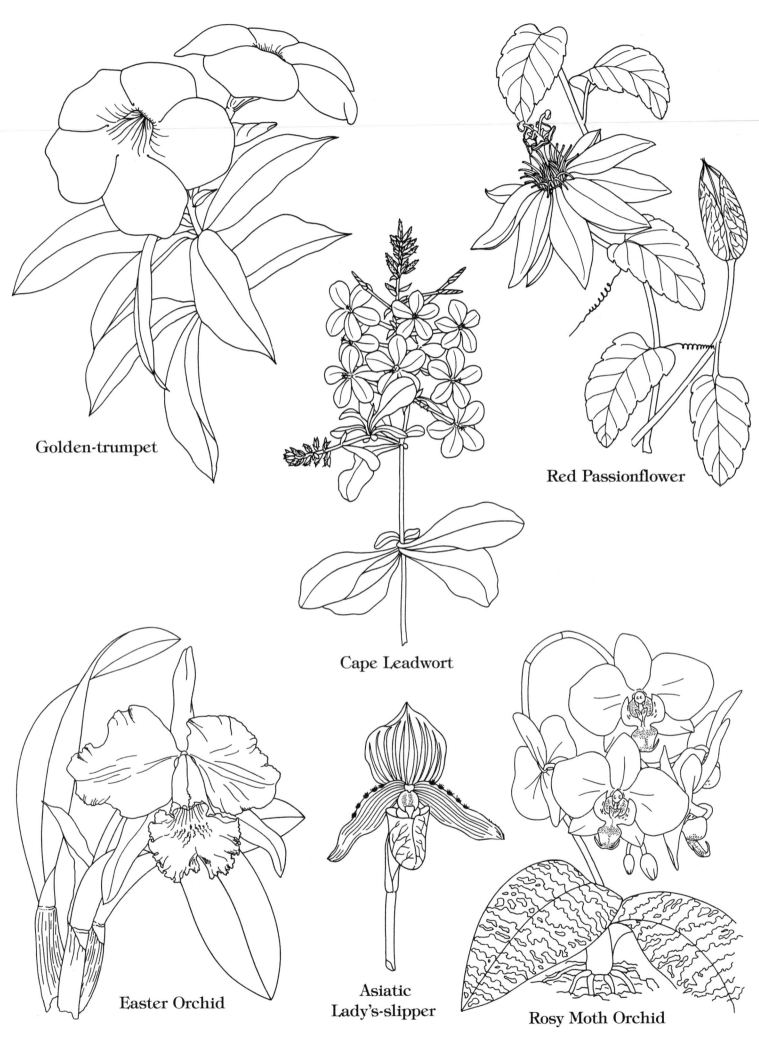

Golden-trumpet

Cape Leadwort

Red Passionflower

Easter Orchid

Asiatic
Lady's-slipper

Rosy Moth Orchid

Pukiawe (Styphelia tameiameiae)

Anthurium

Beach Naupaka

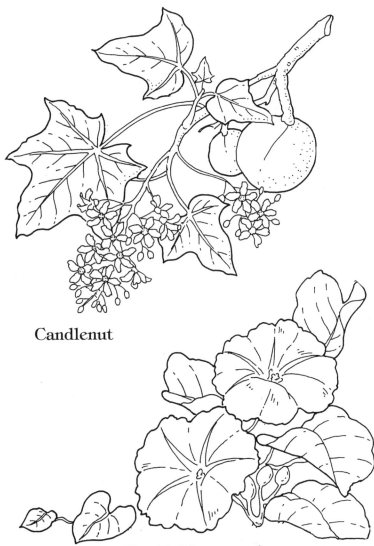

Candlenut

Red Ginger

Beach Morning-glory

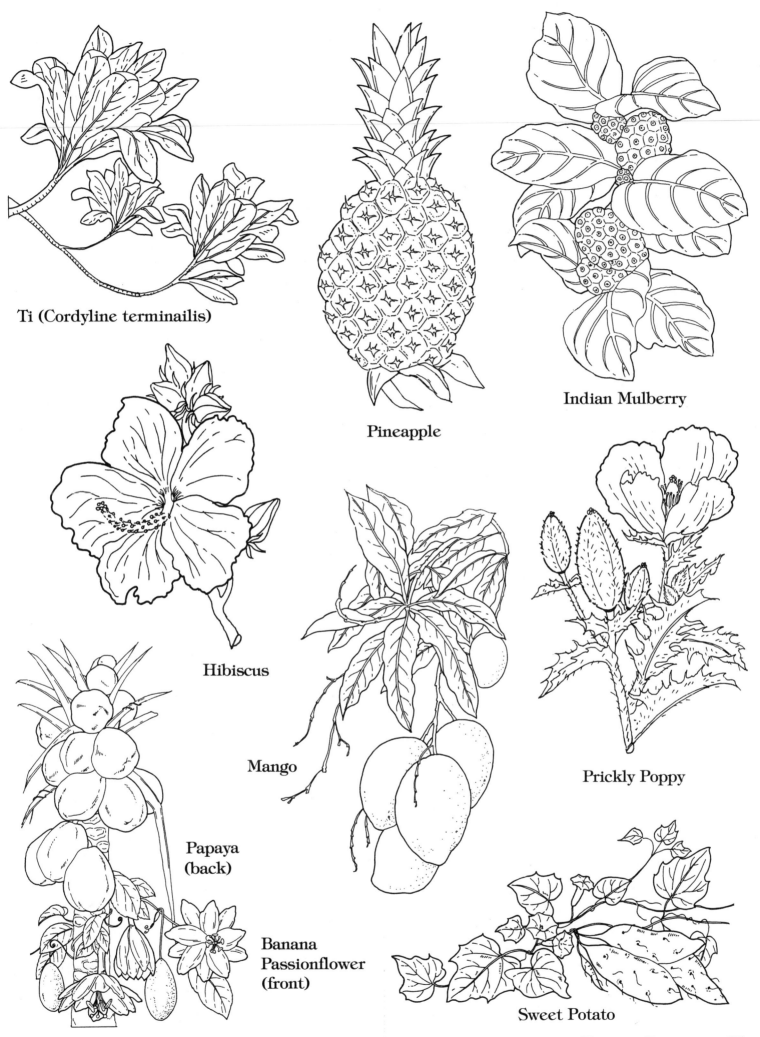

Ti (Cordyline terminailis)

Pineapple

Indian Mulberry

Hibiscus

Mango

Prickly Poppy

Papaya
(back)

Banana
Passionflower
(front)

Sweet Potato

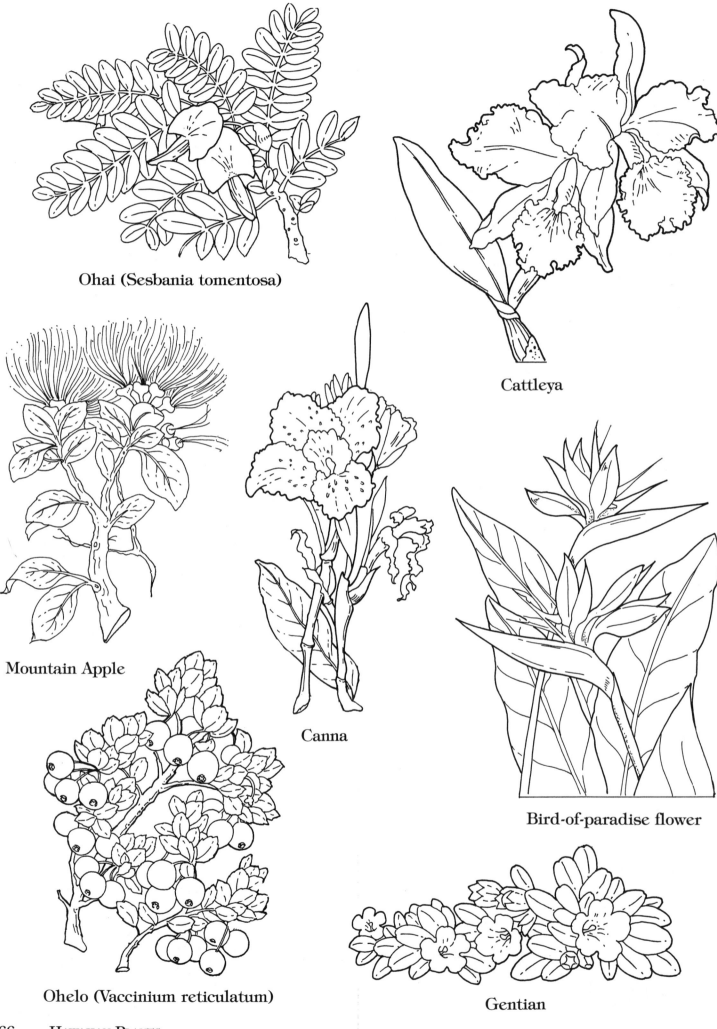

Ohai (Sesbania tomentosa)

Cattleya

Mountain Apple

Canna

Bird-of-paradise flower

Ohelo (Vaccinium reticulatum)

Gentian

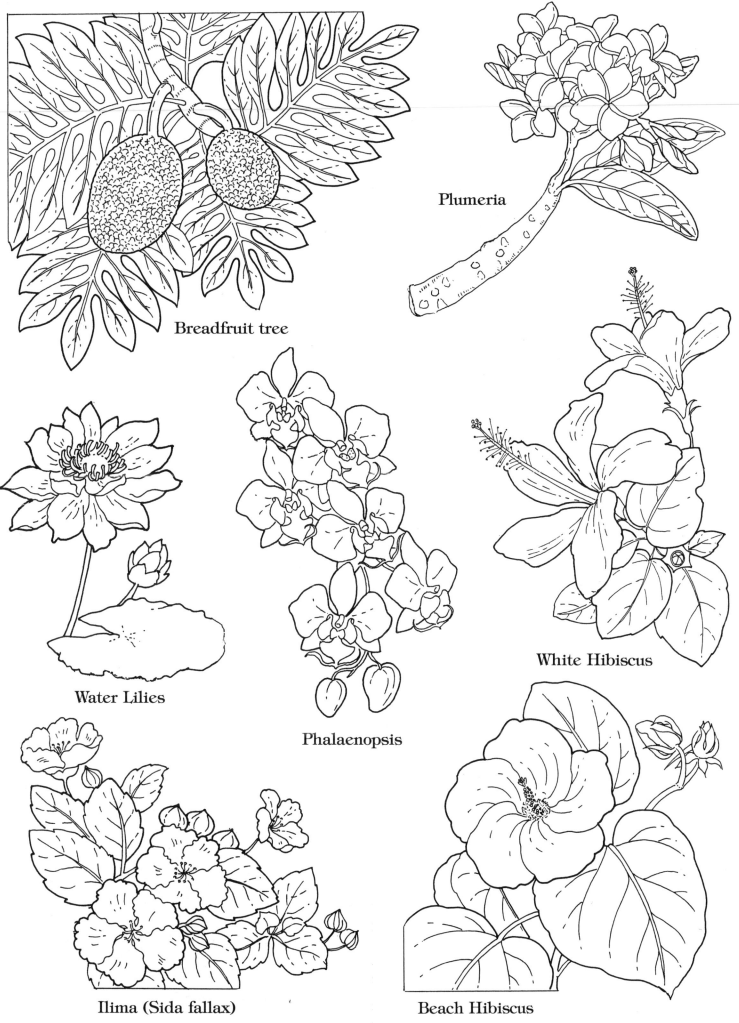

Breadfruit tree

Plumeria

Water Lilies

Phalaenopsis

White Hibiscus

Ilima (Sida fallax)

Beach Hibiscus

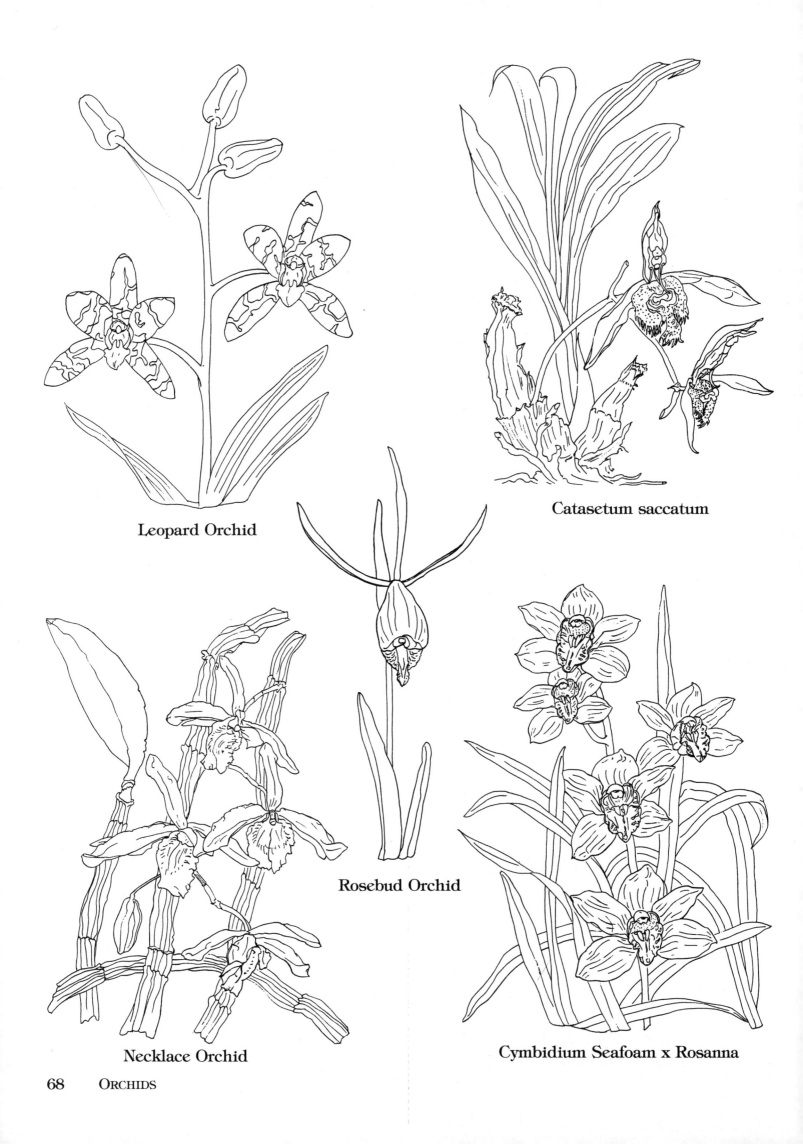

Leopard Orchid

Catasetum saccatum

Rosebud Orchid

Necklace Orchid

Cymbidium Seafoam x Rosanna

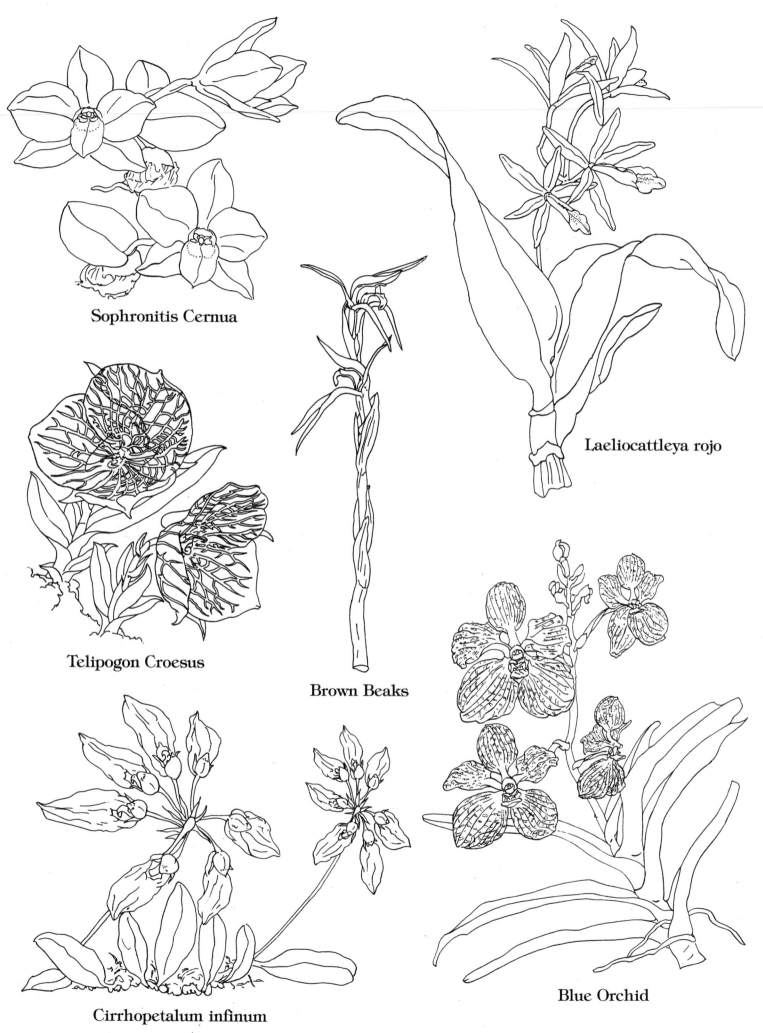

Sophronitis Cernua

Telipogon Croesus

Brown Beaks

Laeliocattleya rojo

Blue Orchid

Cirrhopetalum infinum

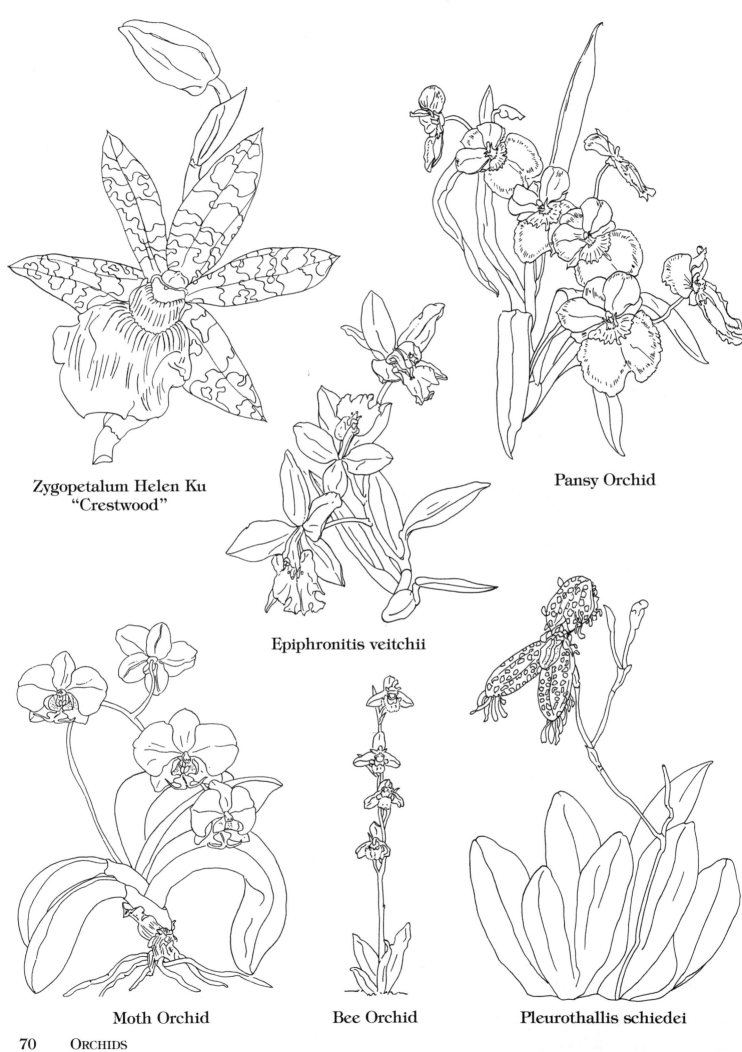

Zygopetalum Helen Ku
"Crestwood"

Pansy Orchid

Epiphronitis veitchii

Moth Orchid

Bee Orchid

Pleurothallis schiedei

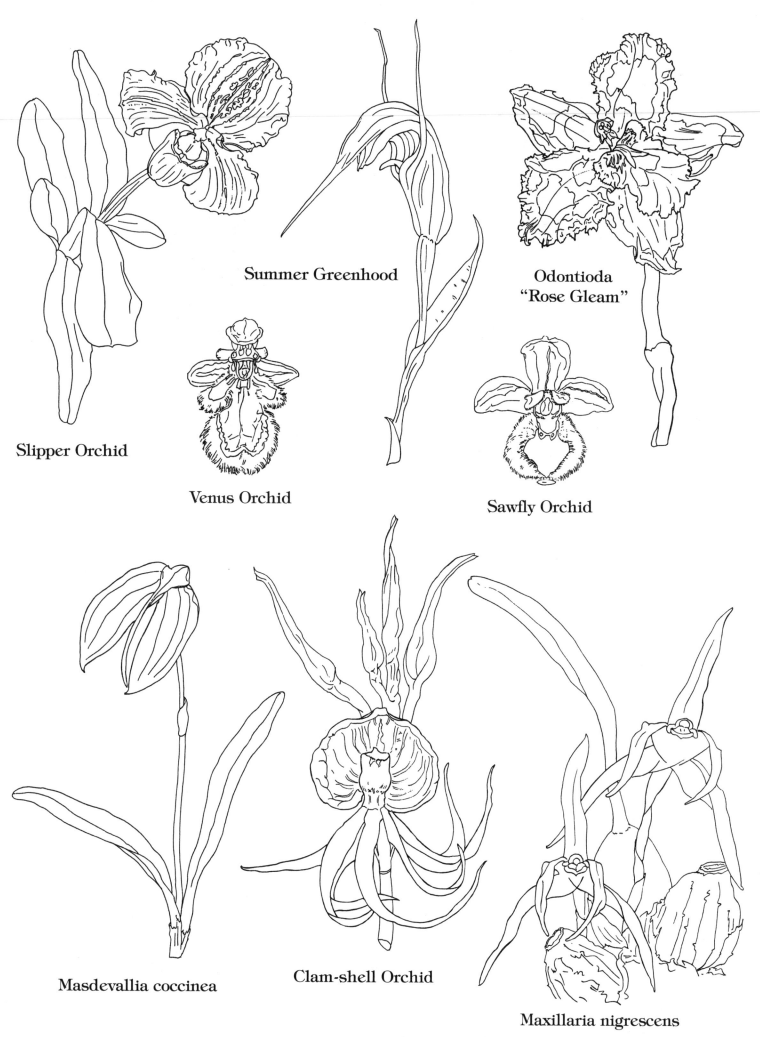

Summer Greenhood

Odontioda
"Rose Gleam"

Slipper Orchid

Venus Orchid

Sawfly Orchid

Masdevallia coccinea

Clam-shell Orchid

Maxillaria nigrescens

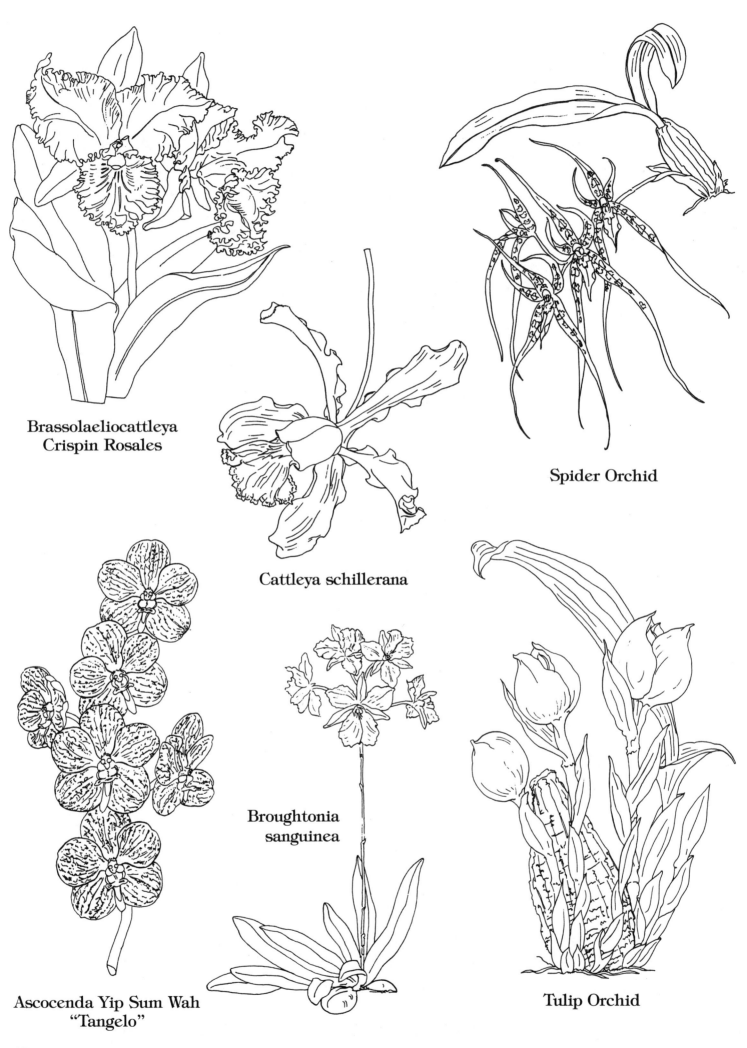

Brassolaeliocattleya
Crispin Rosales

Cattleya schillerana

Spider Orchid

Ascocenda Yip Sum Wah
"Tangelo"

Broughtonia
sanguinea

Tulip Orchid

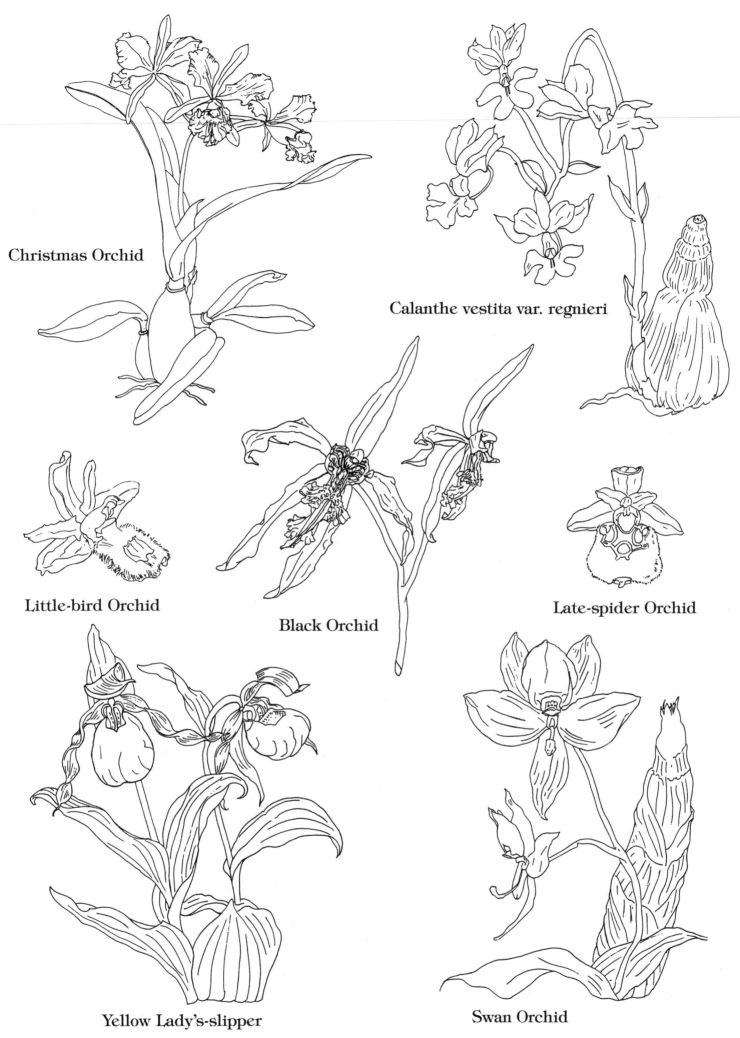

Christmas Orchid

Calanthe vestita var. regnieri

Little-bird Orchid

Black Orchid

Late-spider Orchid

Yellow Lady's-slipper

Swan Orchid

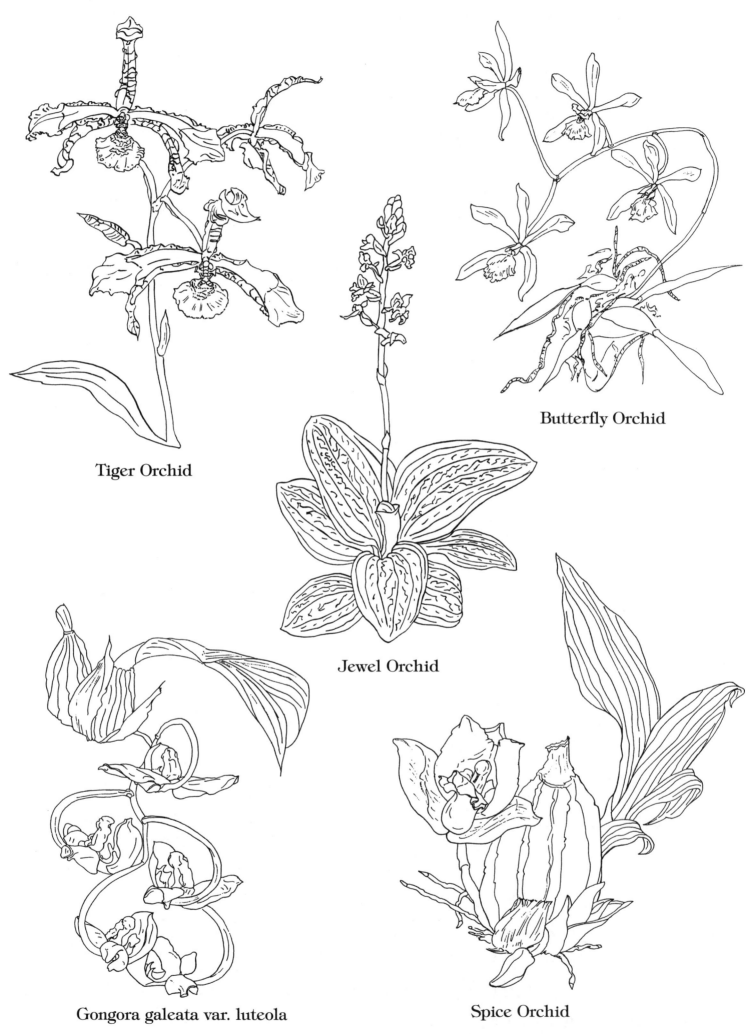

Tiger Orchid

Butterfly Orchid

Jewel Orchid

Gongora galeata var. luteola

Spice Orchid

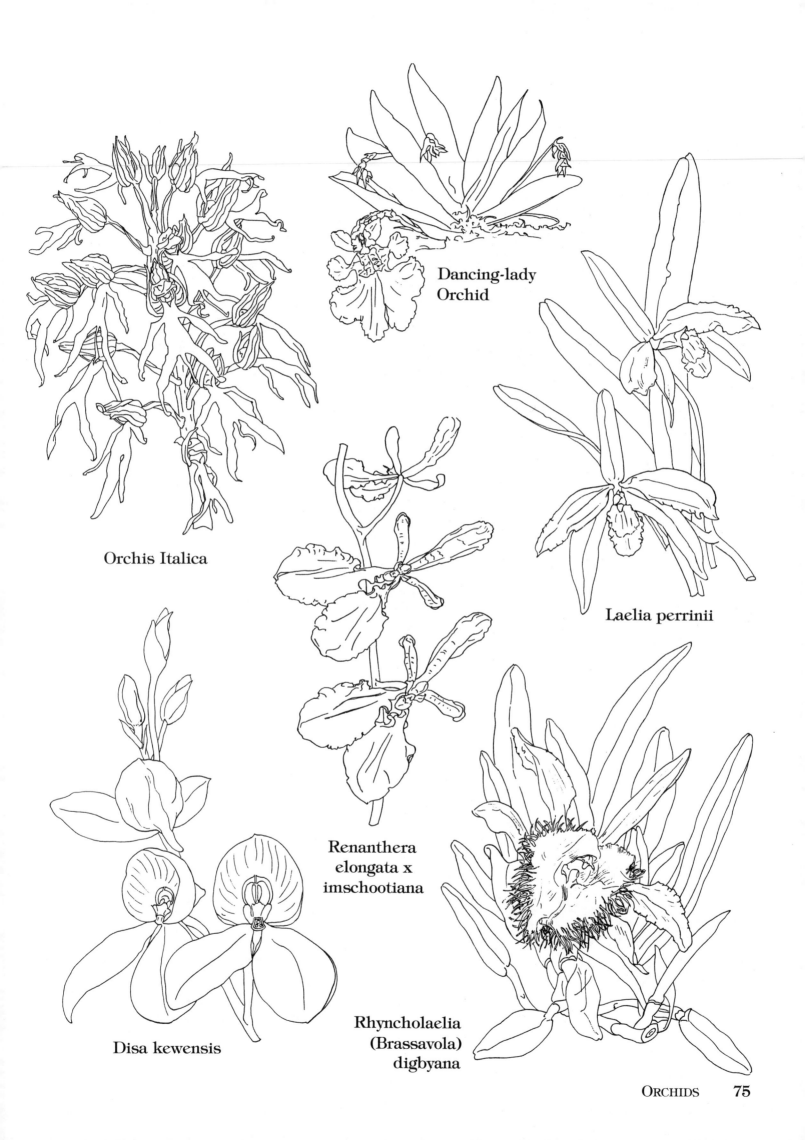

Dancing-lady
Orchid

Orchis Italica

Laelia perrinii

Renanthera
elongata x
imschootiana

Disa kewensis

Rhyncholaelia
(Brassavola)
digbyana

Cecile Brunner (Polylantha)

Brandy (Hybrid Tea)

Cinderella (Miniature)

Color Magic
(Hybrid Tea)

America (Climber)

China Doll (Polyantha)

Hermosa (China)

Sweet Briar Rose (species rose)

First Edition (Floribunda)

Holy Toledo (Miniature)

Bahia (Floribunda)

Gold Coin (Miniature)

Common Moss (Moss)

First Prize
(Hybrid Tea)

Garden Party
(Hybrid Tea)

Double Delight (Hybrid Tea)

Golden Showers (Climber)

Great Maiden's Blush (Alba)

Duchesse de Brabant (Tea)

Europeana (Floribunda)

York and Lancaster (Damask)

Royal Highness (Hybrid Tea)

Pink Parfait (Grandiflora)

Peace (Hybrid Tea)

Mister Lincoln
(Hybrid Tea)

Maréchal Niel (Noisette)

Arizona (Grandiflora)

New Dawn (Climber)

Blaze (Climber)

Sundowner (Grandiflora)

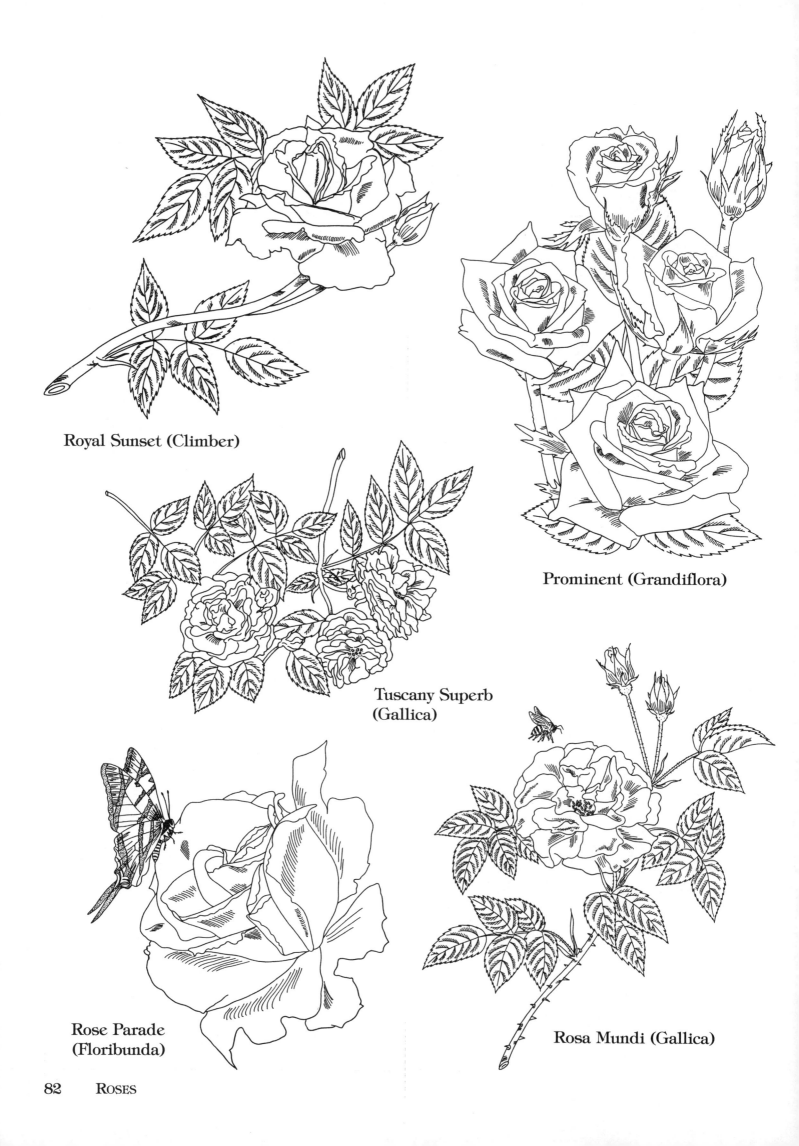

Royal Sunset (Climber)

Prominent (Grandiflora)

Tuscany Superb
(Gallica)

Rose Parade
(Floribunda)

Rosa Mundi (Gallica)

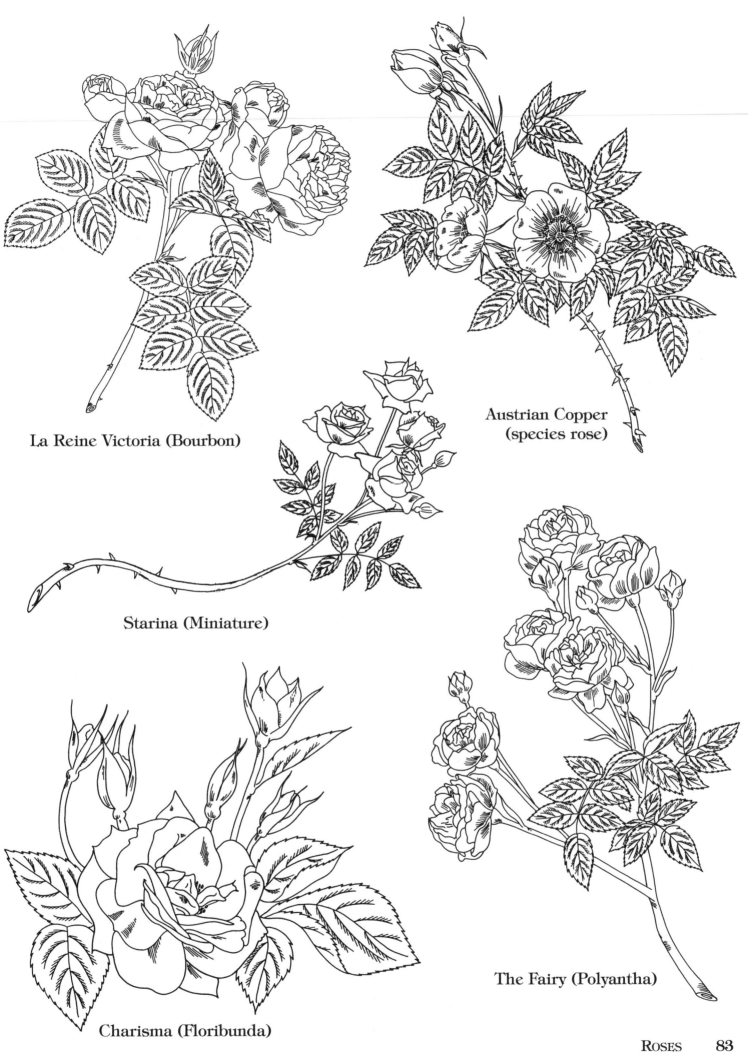

La Reine Victoria (Bourbon)

Austrian Copper
(species rose)

Starina (Miniature)

The Fairy (Polyantha)

Charisma (Floribunda)

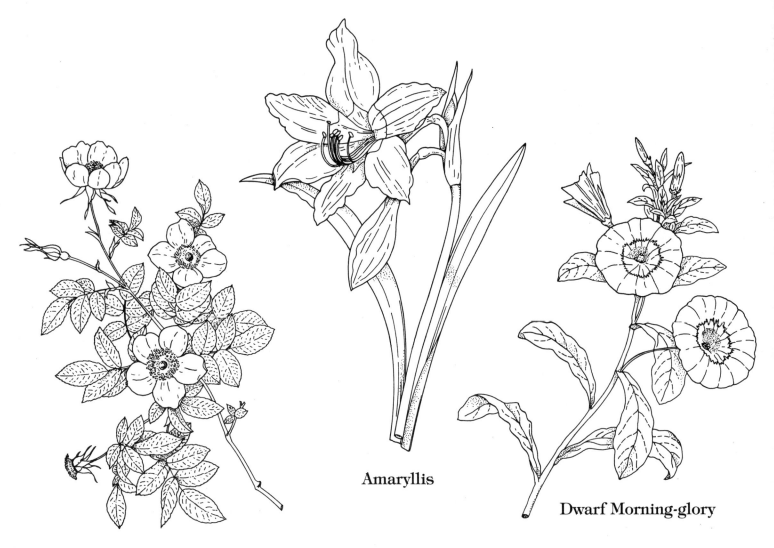

Amaryllis

Dwarf Morning-glory

Alpine Briar Rose

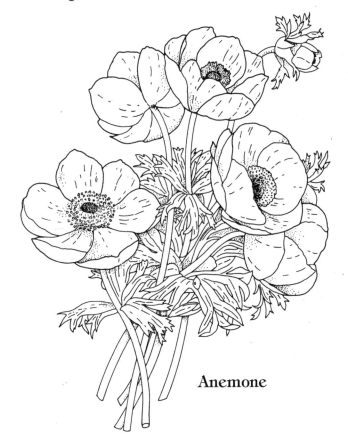

Anemone

China Aster

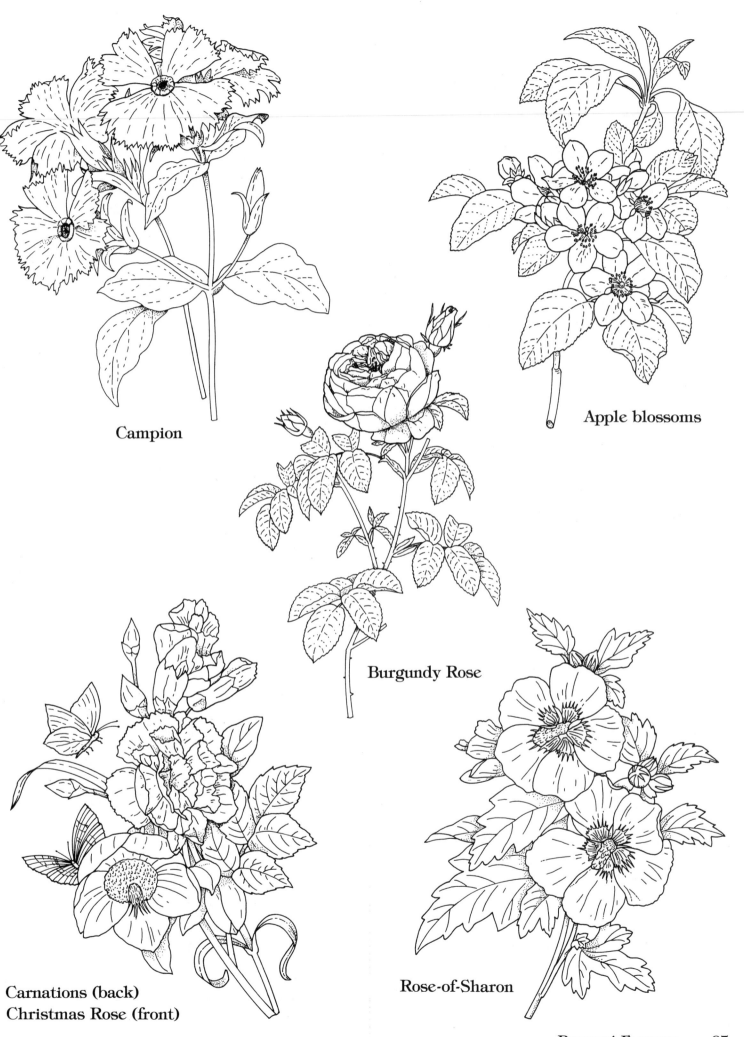

Campion

Apple blossoms

Burgundy Rose

Carnations (back)
Christmas Rose (front)

Rose-of-Sharon

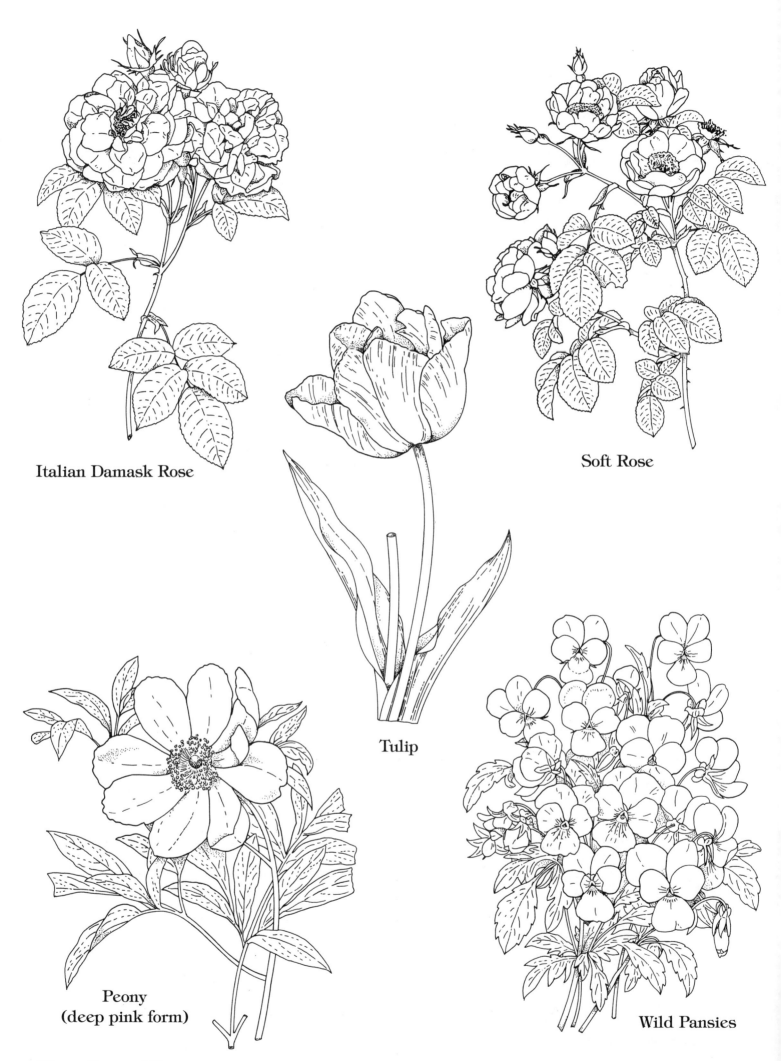

Italian Damask Rose

Soft Rose

Tulip

Peony
(deep pink form)

Wild Pansies

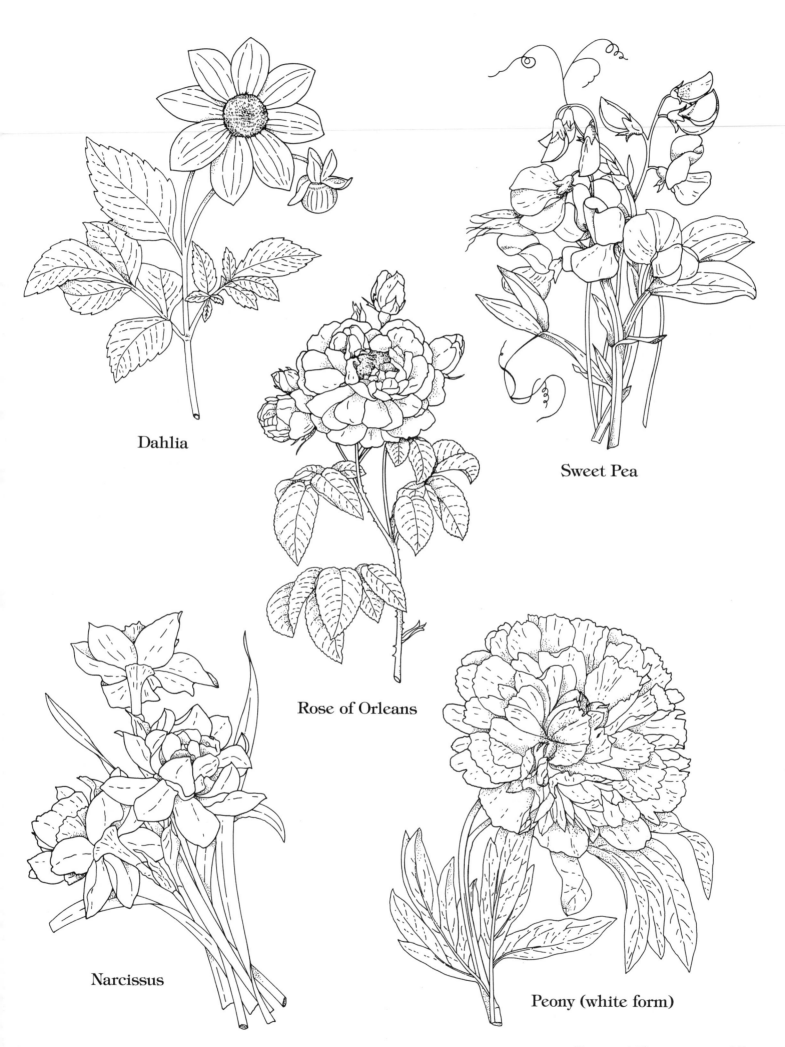

Dahlia

Sweet Pea

Rose of Orleans

Narcissus

Peony (white form)

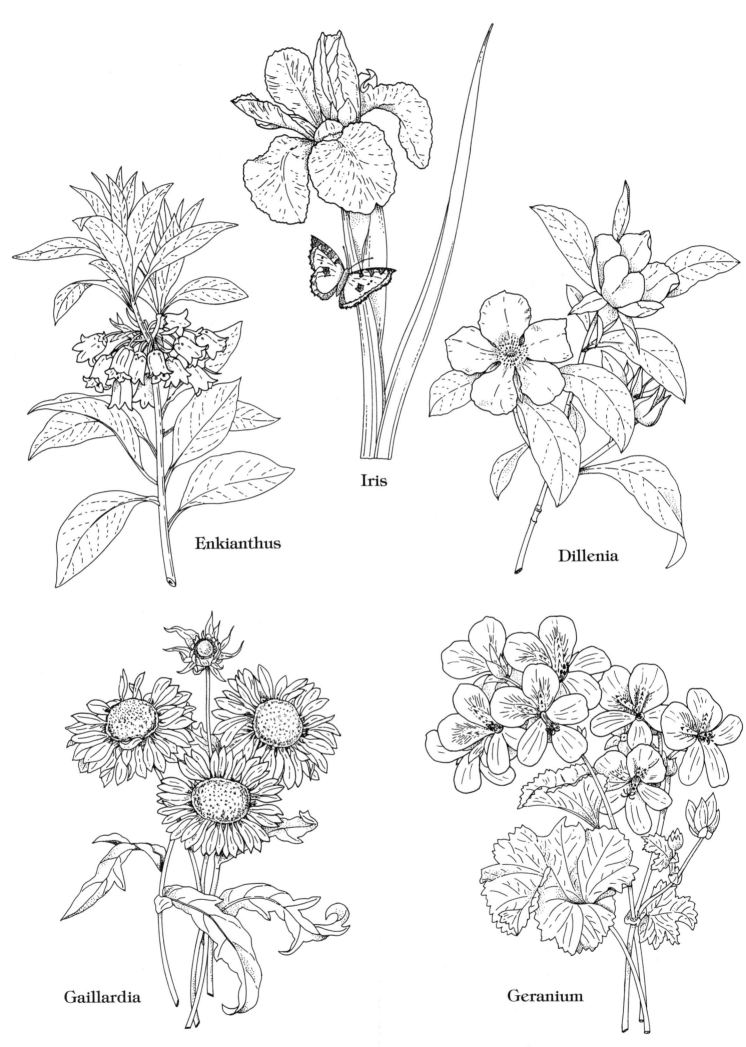

Enkianthus

Iris

Dillenia

Gaillardia

Geranium

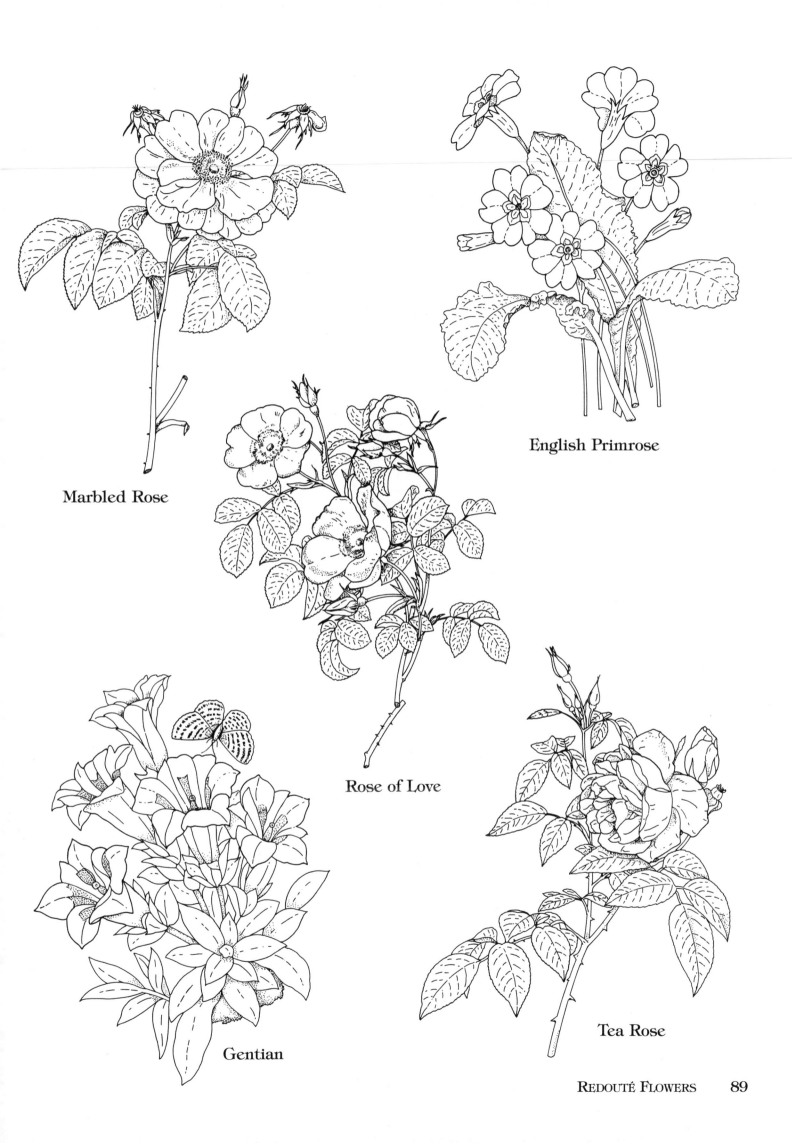

Marbled Rose

English Primrose

Rose of Love

Gentian

Tea Rose

Bolete

Larch Lover

Death Cap

King Bolete

Rough-stemmed Boletes

Pine Mushroom

Boletes

MUSHROOMS

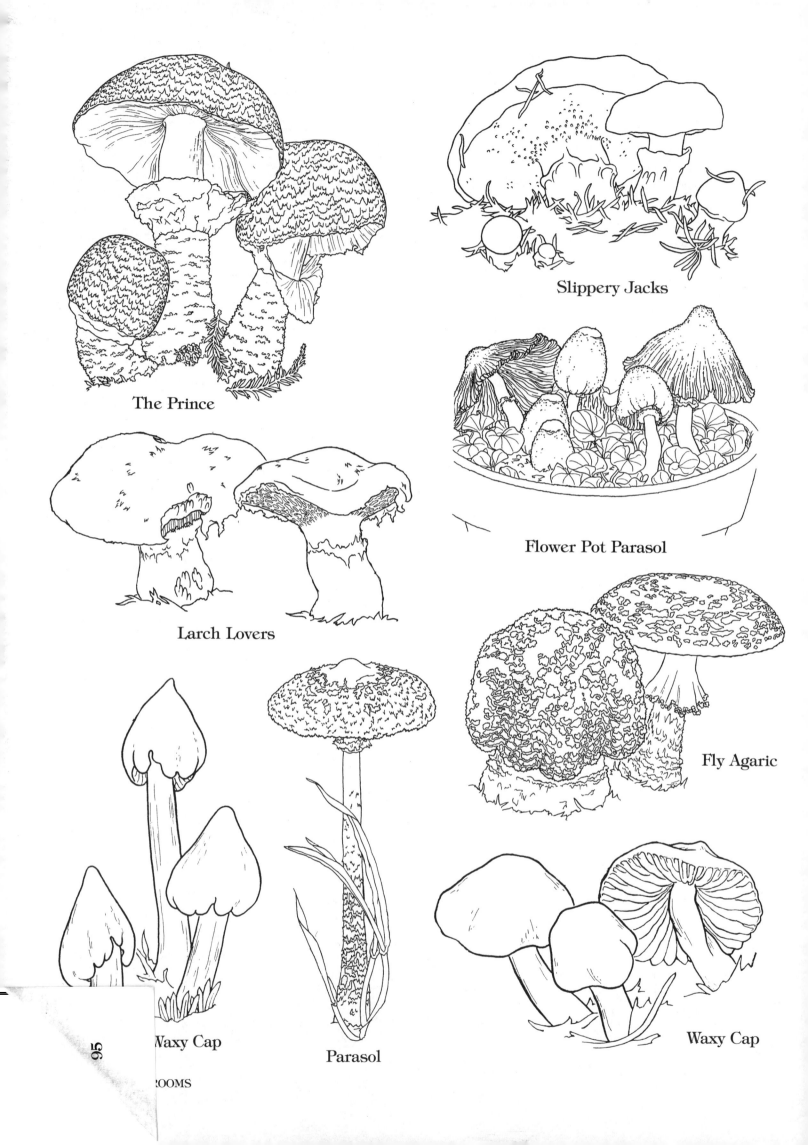

The Prince

Slippery Jacks

Larch Lovers

Flower Pot Parasol

Fly Agaric

Waxy Cap

Parasol

Waxy Cap

ROOMS

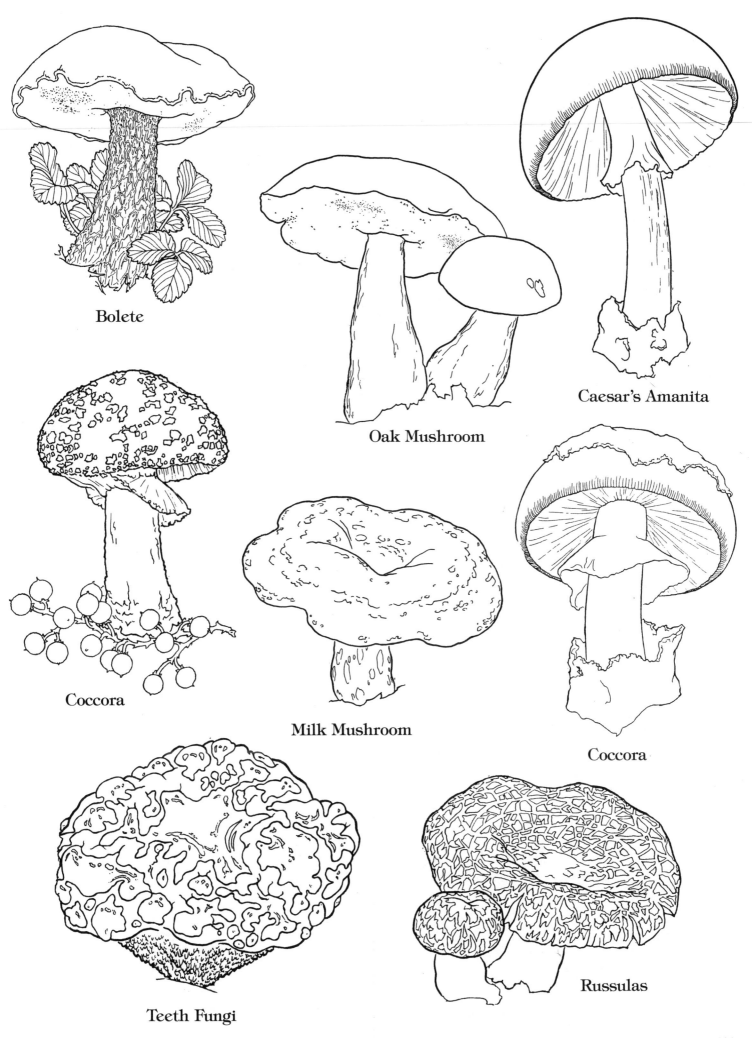

Bolete

Oak Mushroom

Caesar's Amanita

Coccora

Milk Mushroom

Coccora

Teeth Fungi

Russulas

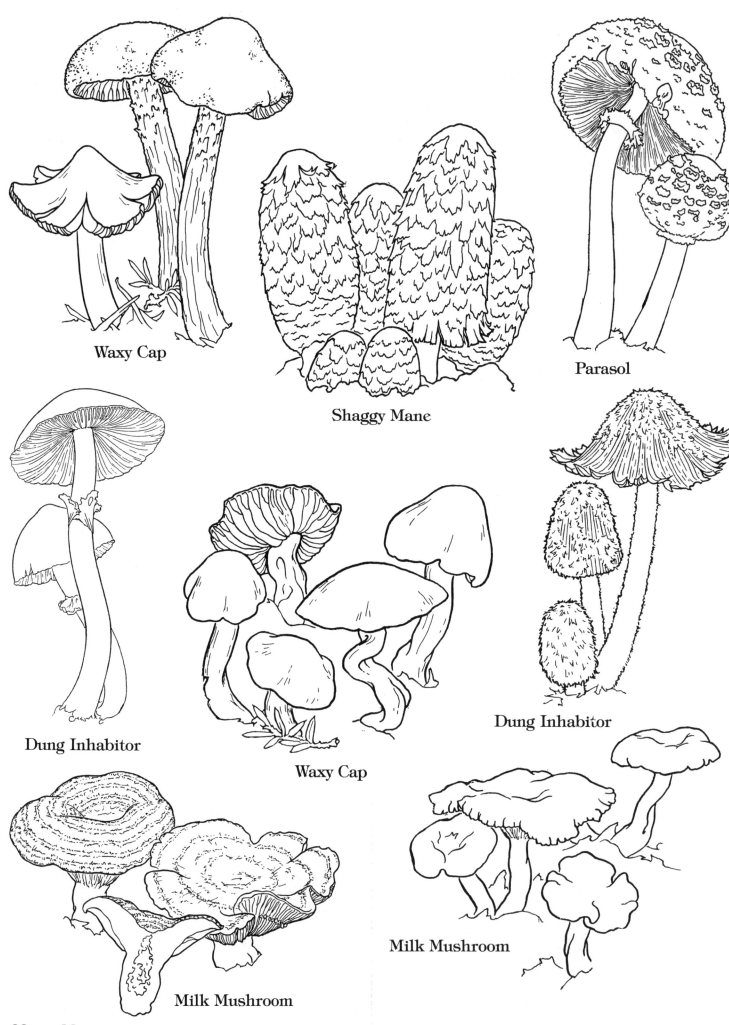

Waxy Cap

Shaggy Mane

Parasol

Dung Inhabitor

Waxy Cap

Dung Inhabitor

Milk Mushroom

Milk Mushroom

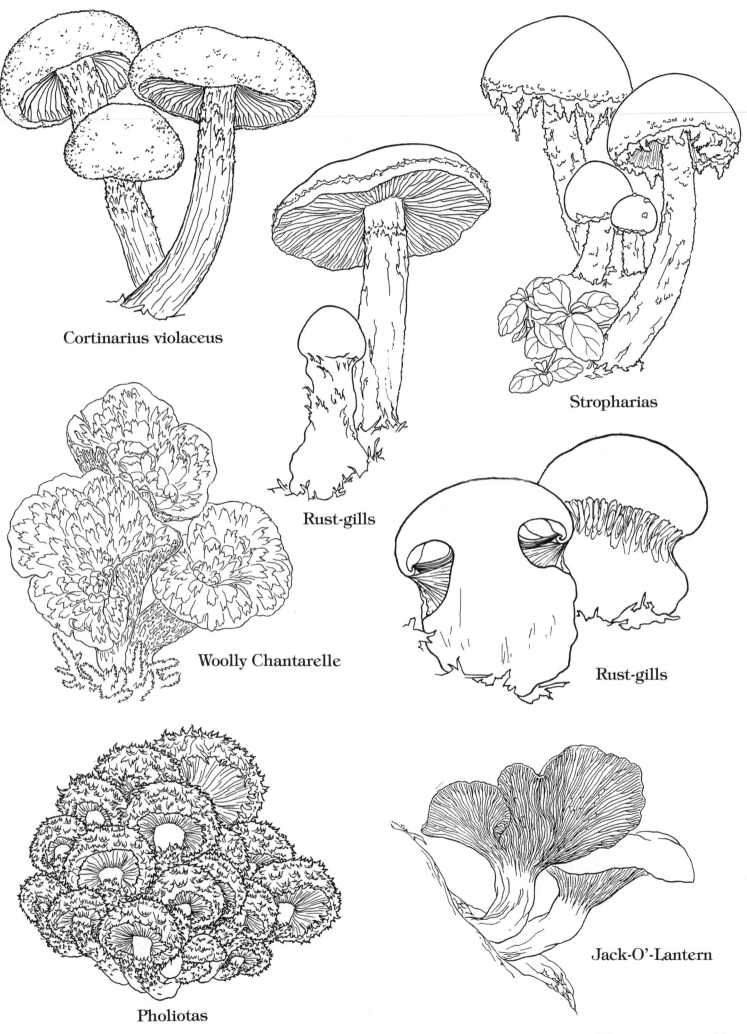

Cortinarius violaceus

Rust-gills

Stropharias

Woolly Chantarelle

Rust-gills

Pholiotas

Jack-O'-Lantern

Pholiotas

Teeth Fungi

Blewitt

Chantarelles

Coral Fungi

Polypores

Coral Fungi

Teeth Fungi

Horn-of-Plenty

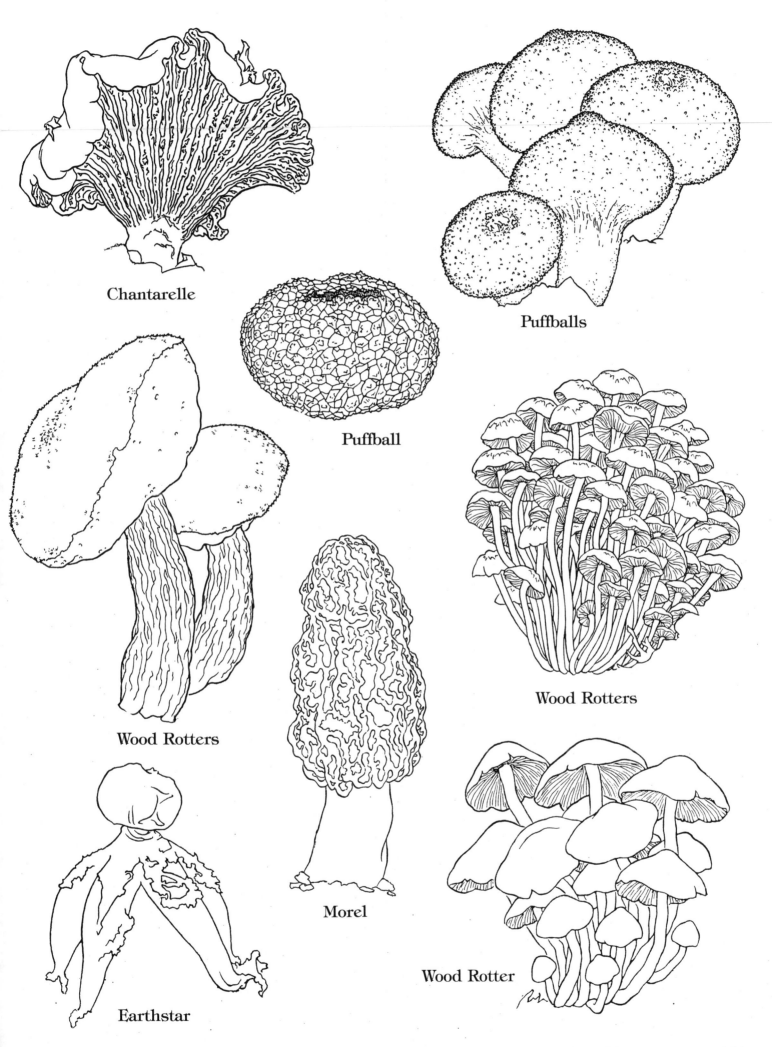

Chantarelle

Puffballs

Puffball

Wood Rotters

Wood Rotters

Morel

Wood Rotter

Earthstar

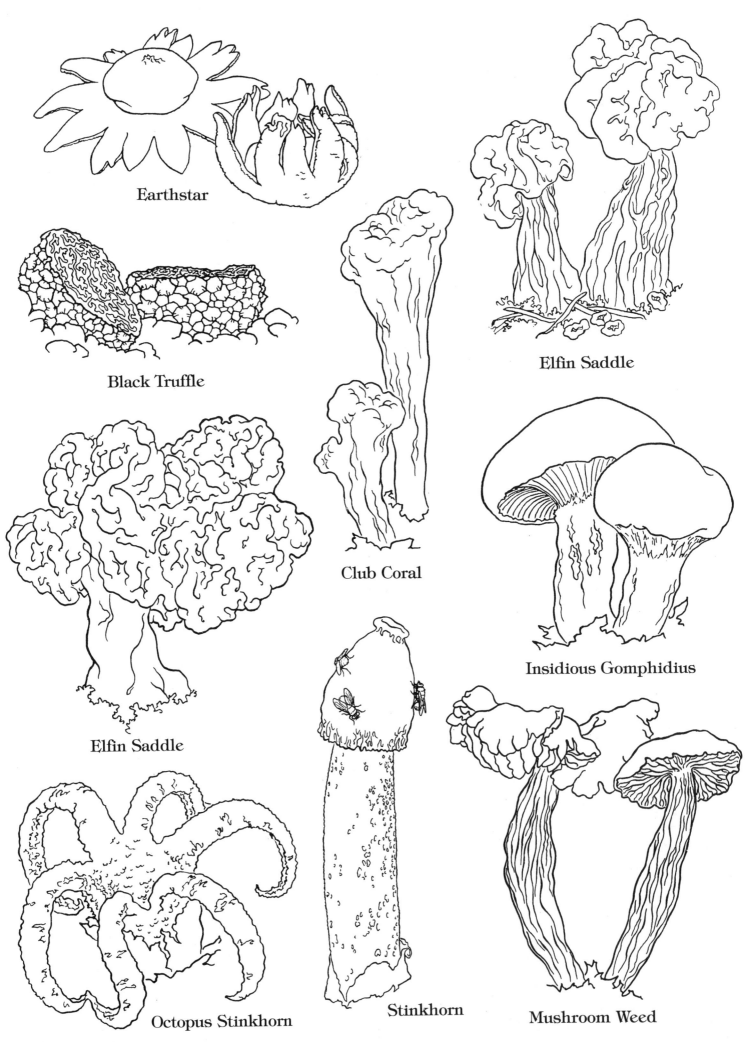

Earthstar

Black Truffle

Elfin Saddle

Club Coral

Elfin Saddle

Insidious Gomphidius

Octopus Stinkhorn

Stinkhorn

Mushroom Weed

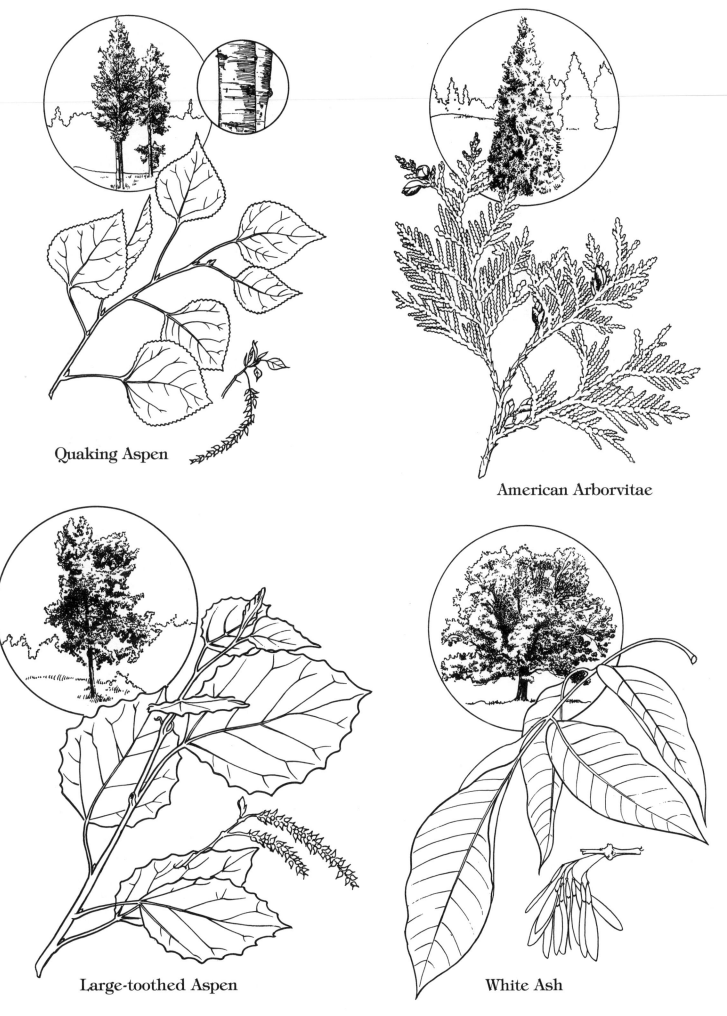

Quaking Aspen

American Arborvitae

Large-toothed Aspen

White Ash

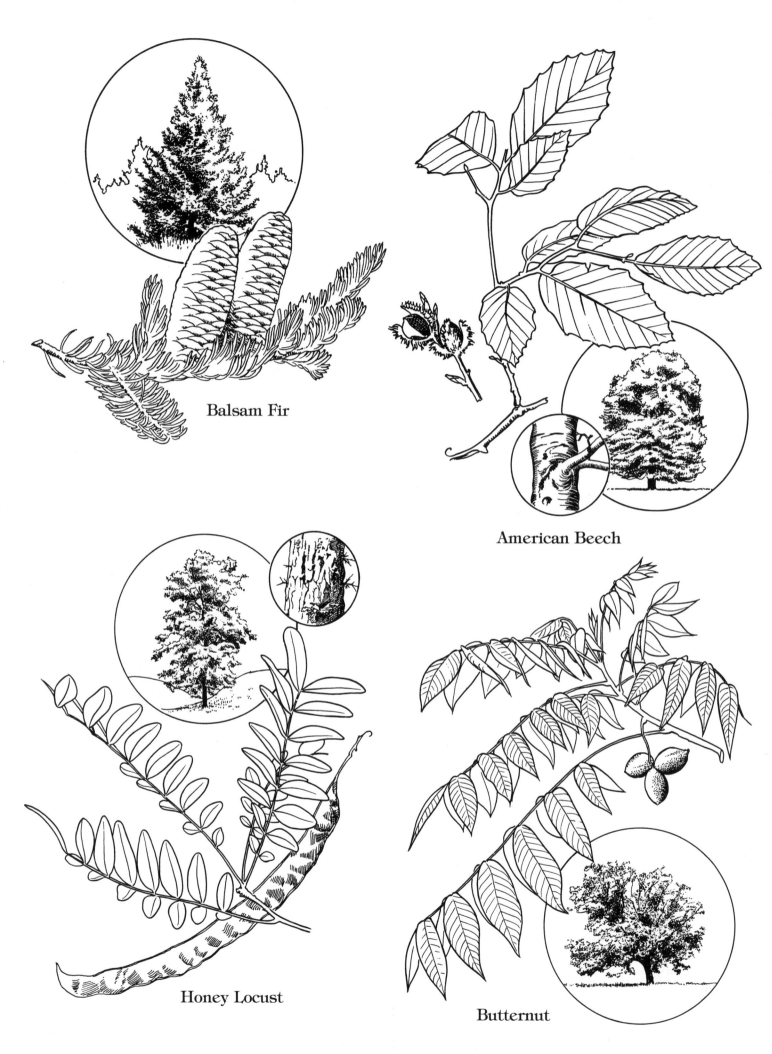

Balsam Fir

American Beech

Honey Locust

Butternut

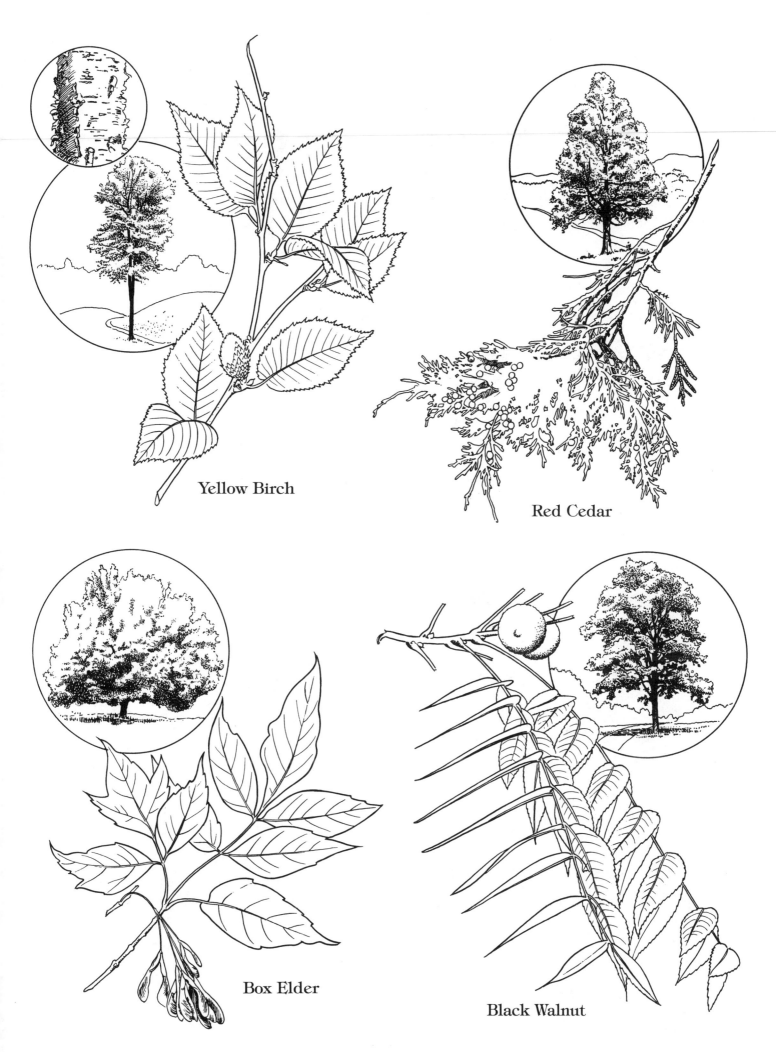

Yellow Birch

Red Cedar

Box Elder

Black Walnut

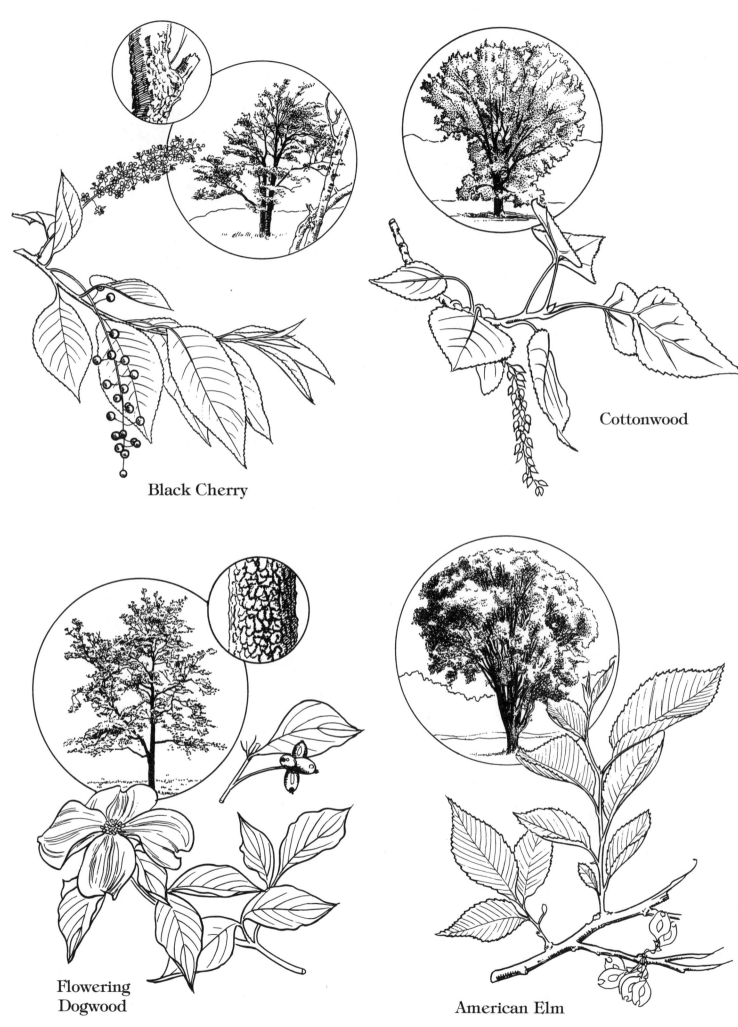

Black Cherry

Cottonwood

Flowering
Dogwood

American Elm

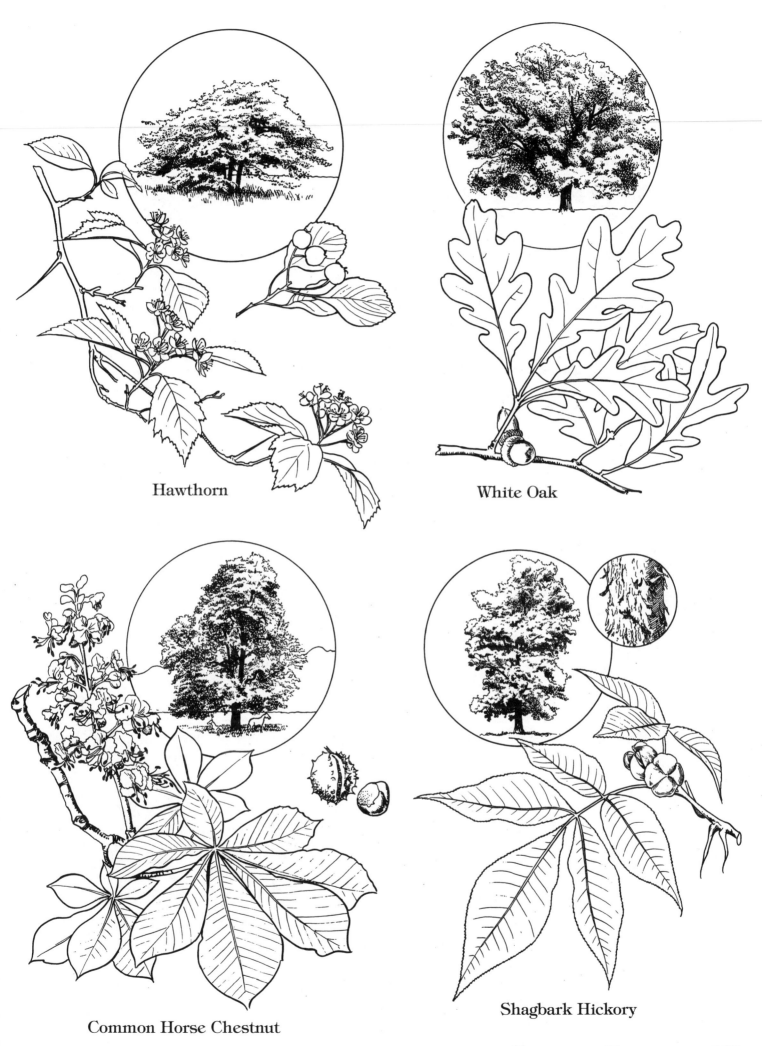

Hawthorn

White Oak

Common Horse Chestnut

Shagbark Hickory

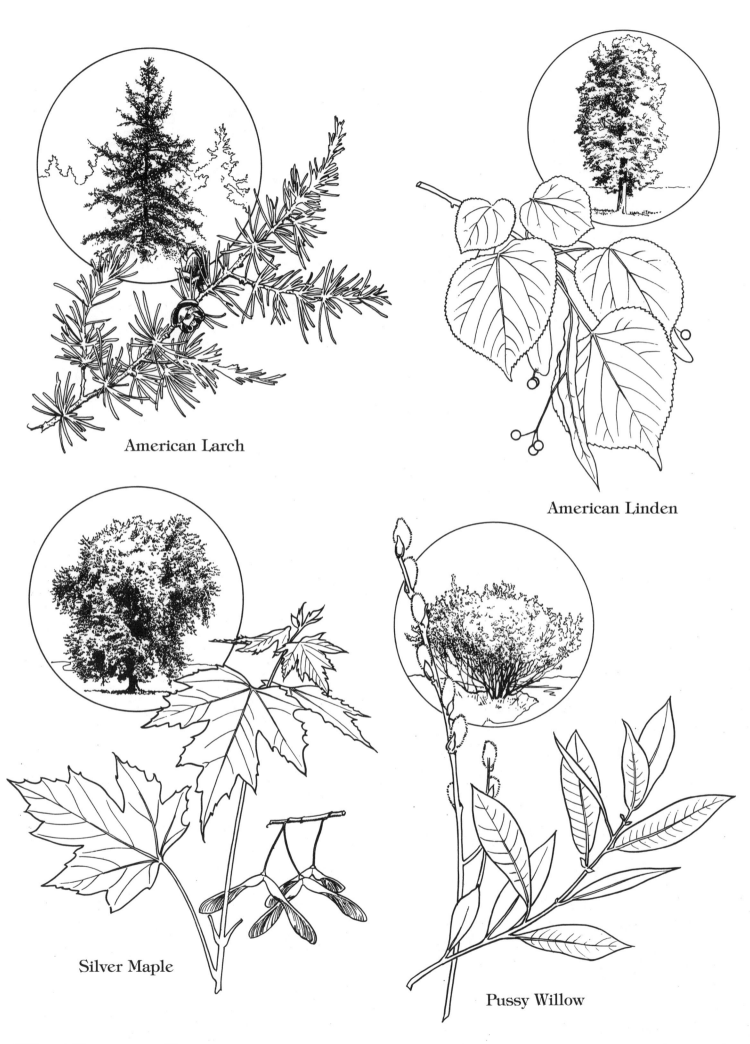

American Larch

American Linden

Silver Maple

Pussy Willow

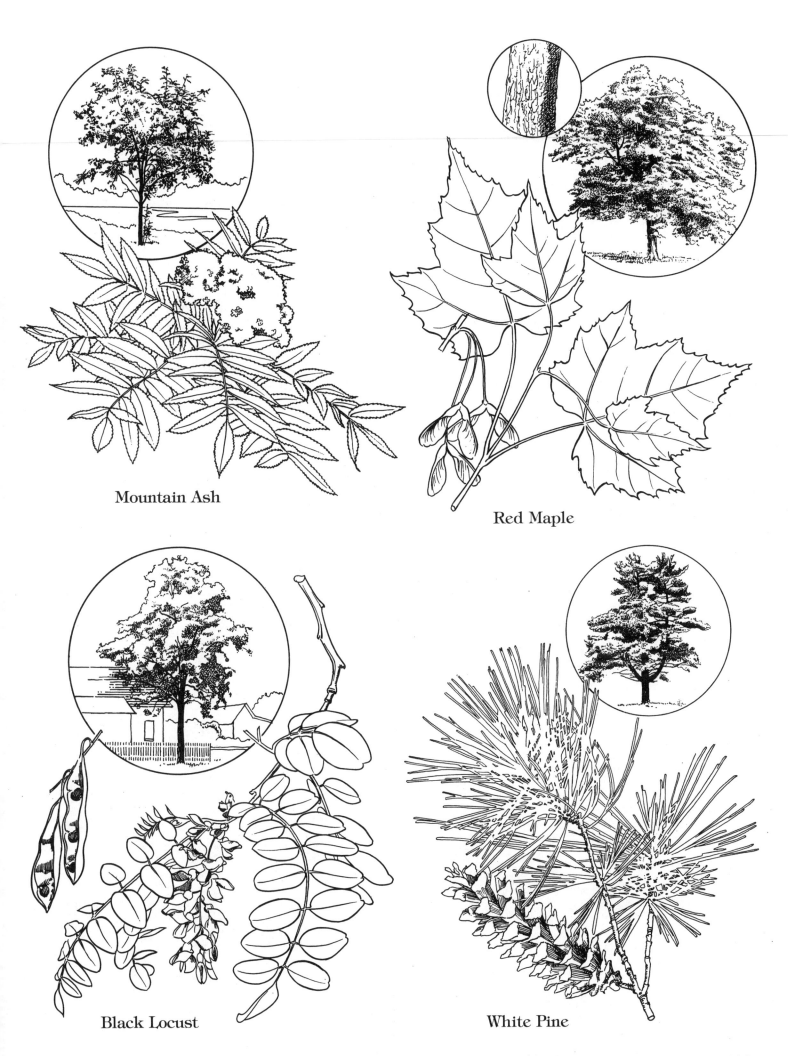

Mountain Ash

Red Maple

Black Locust

White Pine

Tree of Heaven

Sassafras

Staghorn Sumac

White Spruce

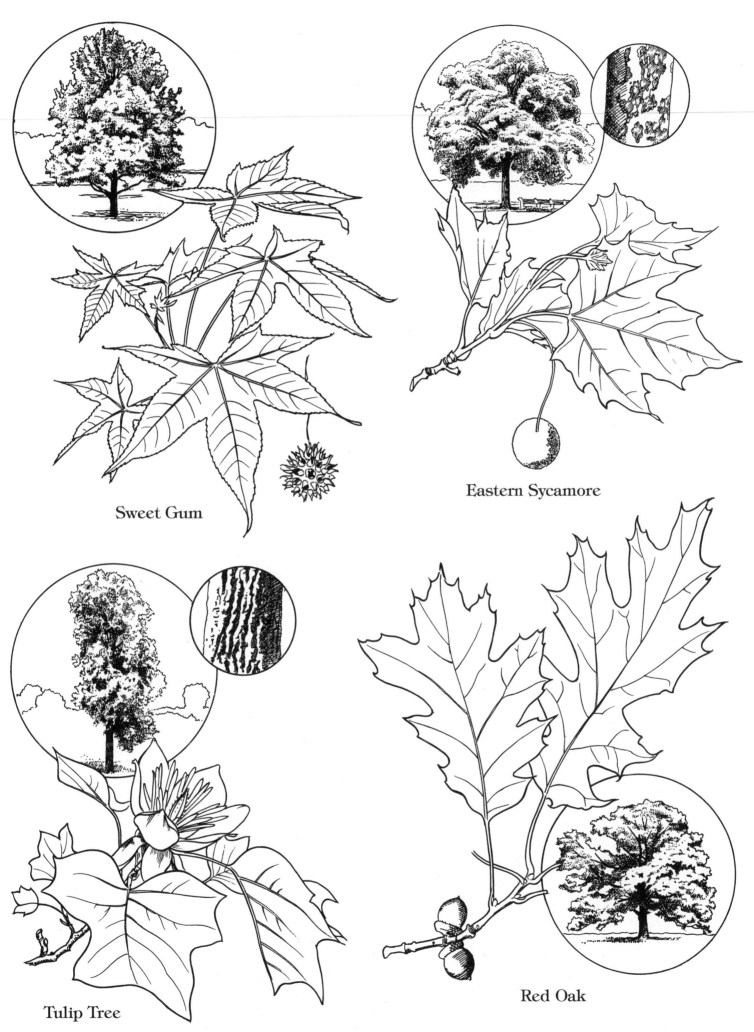

Sweet Gum

Eastern Sycamore

Tulip Tree

Red Oak

Carnation
(State Flower of Ohio)

Mock Orange
(State Flower of Idaho)

Rose
(State Flower of New York,
as well as the National Flower)

Texas Bluebonnet
(State Flower of Texas)

Peony
(State Flower of Indiana)

Indian Paintbrush
(State Flower of Wyoming)

Pasqueflower
(State Flower of South Dakota)

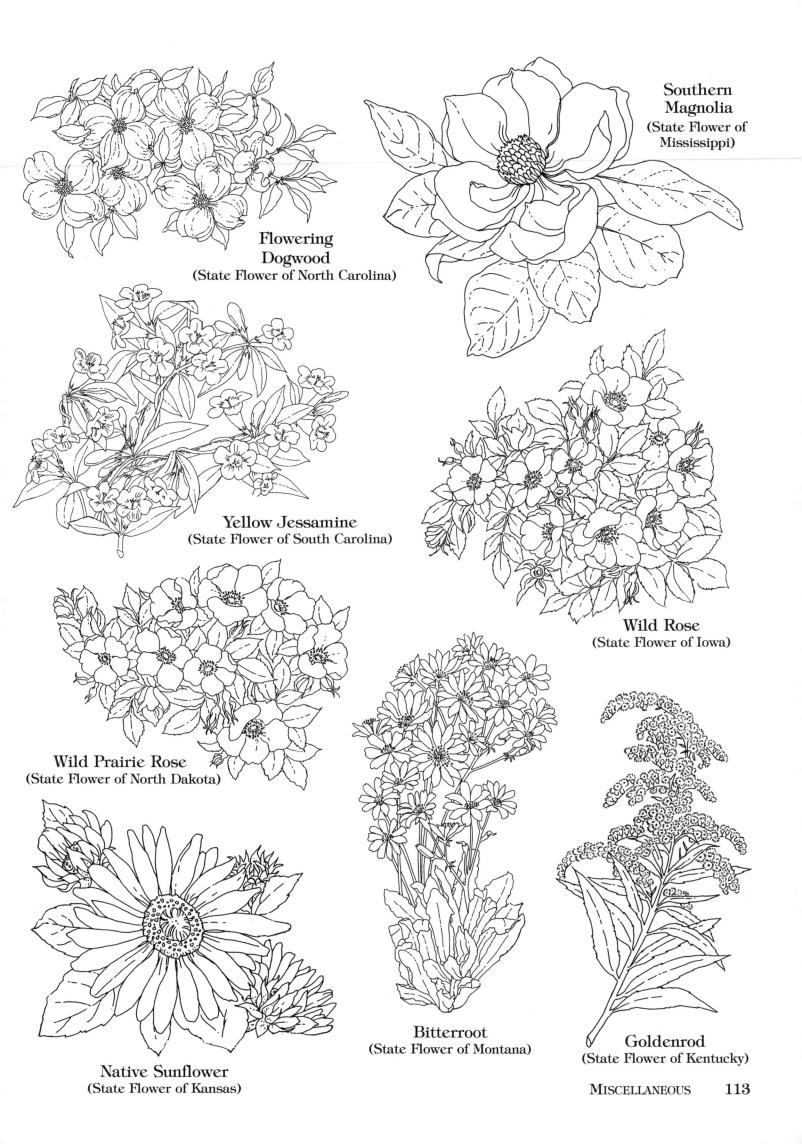

Flowering
Dogwood
(State Flower of North Carolina)

Southern
Magnolia
(State Flower of
Mississippi)

Yellow Jessamine
(State Flower of South Carolina)

Wild Rose
(State Flower of Iowa)

Wild Prairie Rose
(State Flower of North Dakota)

Bitterroot
(State Flower of Montana)

Goldenrod
(State Flower of Kentucky)

Native Sunflower
(State Flower of Kansas)

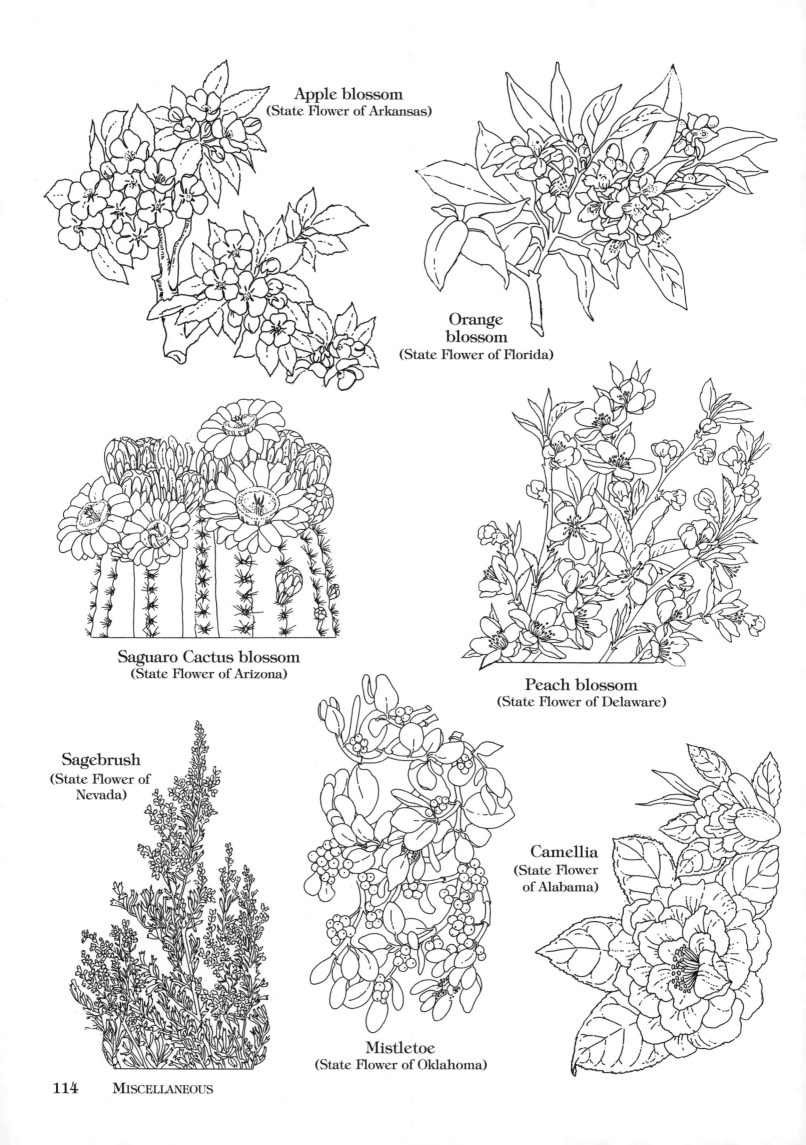

Apple blossom
(State Flower of Arkansas)

Orange
blossom
(State Flower of Florida)

Saguaro Cactus blossom
(State Flower of Arizona)

Peach blossom
(State Flower of Delaware)

Sagebrush
(State Flower of
Nevada)

Mistletoe
(State Flower of Oklahoma)

Camellia
(State Flower
of Alabama)

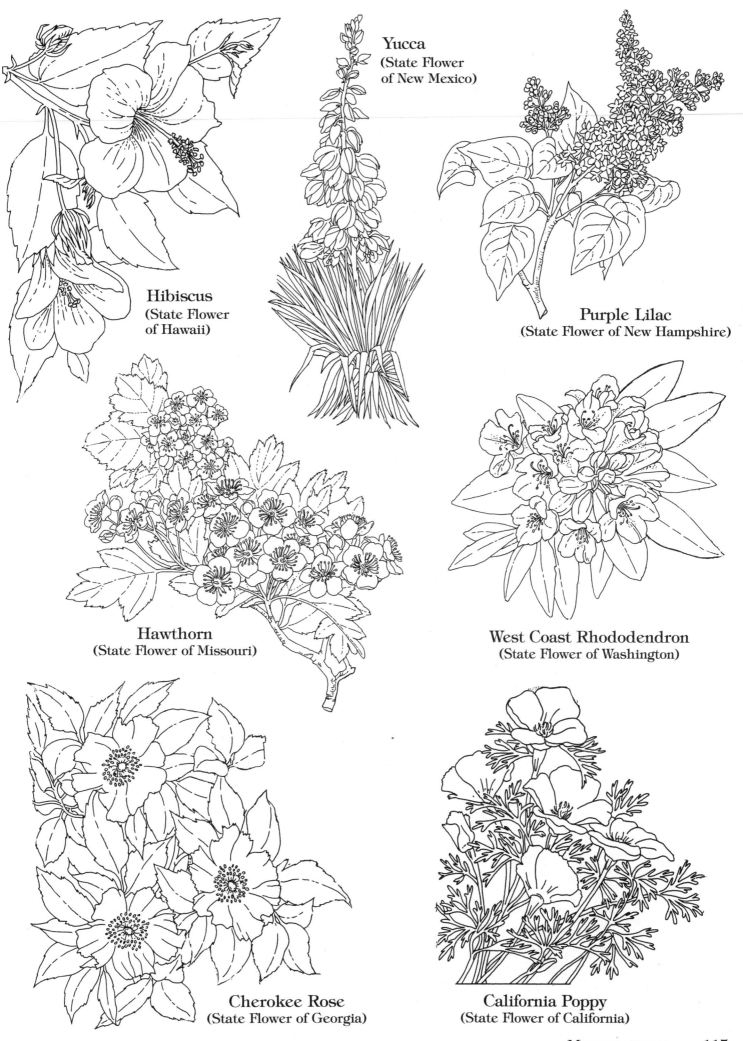

Hibiscus
(State Flower
of Hawaii)

Yucca
(State Flower
of New Mexico)

Purple Lilac
(State Flower of New Hampshire)

Hawthorn
(State Flower of Missouri)

West Coast Rhododendron
(State Flower of Washington)

Cherokee Rose
(State Flower of Georgia)

California Poppy
(State Flower of California)

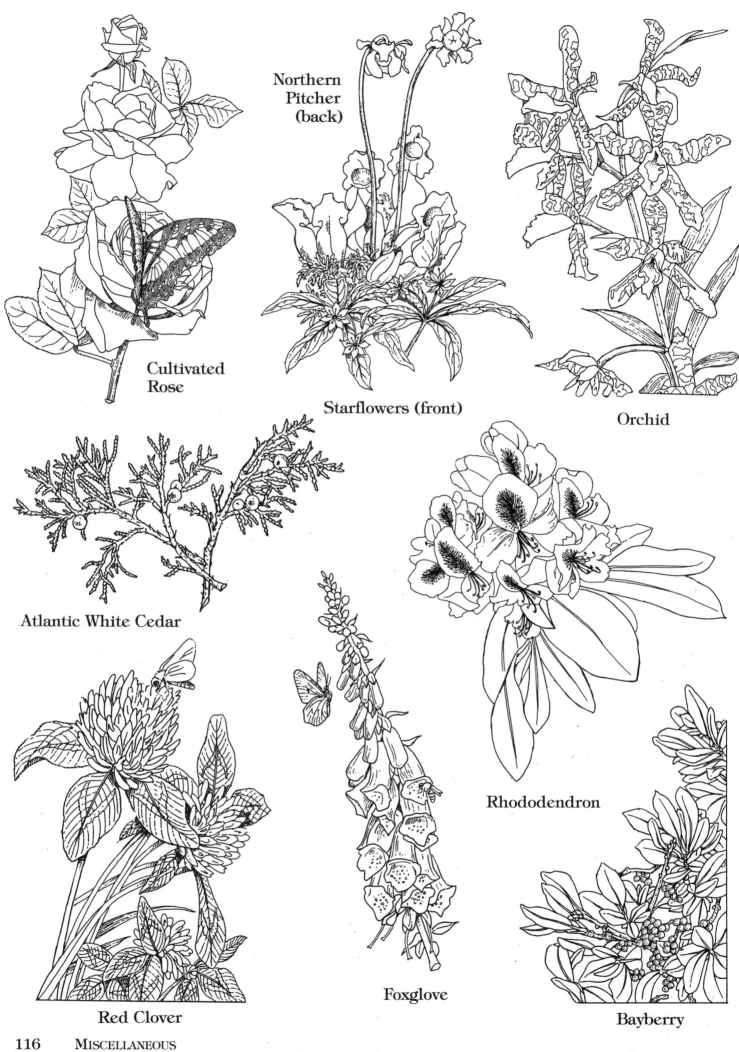

Cultivated
Rose

Northern
Pitcher
(back)

Starflowers (front)

Orchid

Atlantic White Cedar

Rhododendron

Red Clover

Foxglove

Bayberry

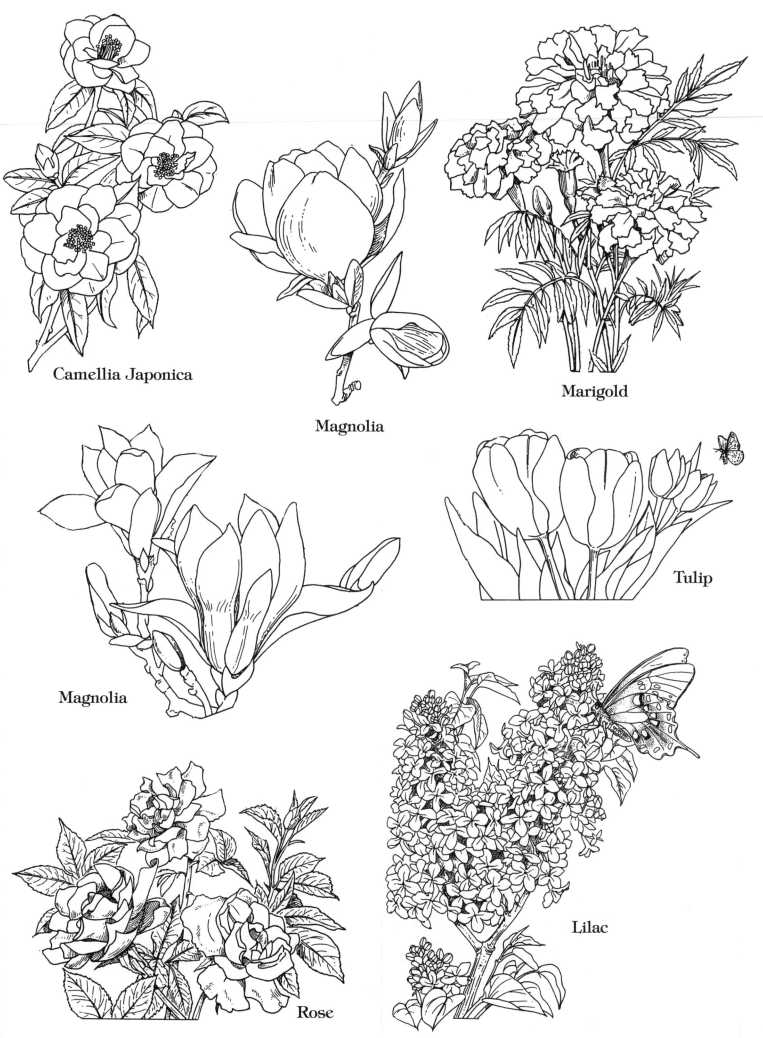

Camellia Japonica

Magnolia

Marigold

Magnolia

Tulip

Rose

Lilac

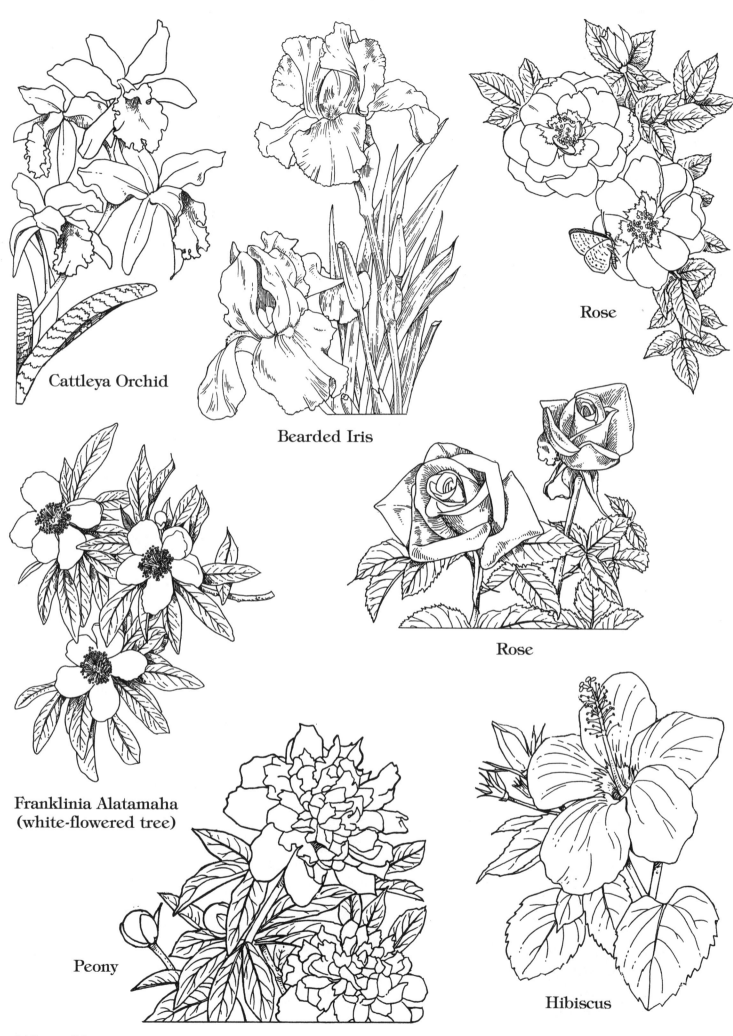

Cattleya Orchid

Bearded Iris

Rose

Rose

Franklinia Alatamaha
(white-flowered tree)

Peony

Hibiscus

Index

(Scientific names only are italicized.)

Sources of the Illustrations

1. *Garden Flowers Coloring Book*, Stefen Bernath (1975)
(Copyright © 1975 by Dover Publications, Inc.) (ISBN 0-436-23142-9)

2. *American Wild Flowers Coloring Book*, Paul E. Kennedy (1971)
(Copyright © 1971 by Dover Publications, Inc.) (ISBN 0-486-20095-7)

   Favorite Wild Flowers Coloring Book, Ilil Arbel (1991)
(Copyright © 1991 by Dover Publications, Inc.) (ISBN 0-486-26729-6)

3. *House Plants Coloring Book*, Stefen Bernath (1976)
(Copyright © 1976 by Dover Publications, Inc.) (ISBN 0-486-23266-2)

4. *Herbs Coloring Book*, Stefen Bernath (1977)
(Copyright © 1977 by Dover Publications, Inc.) (ISBN 0-486-23499-1)

5. *Medicinal Plants Coloring Book*, Ilil Arbel (1993)
(Copyright © 1992 by Dover Publications, Inc.) (ISBN 0-486-27462-4)

6. *Common Weeds Coloring Book*, Stefen Bernath (1976)
(Copyright © 1976 by Dover Publications, Inc.) (ISBN 0-486-23308-1)

7. *The Cactus Coloring Book*, Stefen Bernath (1981)
(Copyright © 1981 by Dover Publications, Inc.) (ISBN 0-486-24097-5)

8. *Tropical Flowers of the World Coloring Book*, Lynda E. Chandler (1981)
(Copyright © 1981 by Dover Publications, Inc.) (ISBN 0-486-24206-4)

9. *Hawaiian Plants and Animals Coloring Book*, Y. S. Green (1998)
(Copyright © 1999 by Dover Publications, Inc.) (ISBN 0-486-40360-2)

10. *Orchids of the World Coloring Book*, Virginia Fowler Elbert (1983)
(Copyright © 1983 by Virginia Fowler Elbert) (ISBN 0-486-24585-3)

11. *Favorite Roses Coloring Book*, Ilil Arbel (1988)
(Copyright © 1988 by Dover Publications, Inc.) (ISBN 0-486-25845-9)

12. *Redouté Flowers Coloring Book*, Charlene Tarbox (1998)
(Copyright © 1998 by Dover Publications, Inc.) (ISBN 0-486-40055-7)

13. *Floral Bouquets*, Charlene Tarbox (1995)
(Copyright © 1995 by Dover Publications, Inc.) (ISBN 0-486-28654-1)

14. *Mushrooms of the World Coloring Book*, Jeannette Bowers (1984)
(Copyright © 1984 by Jeannette Bowers & David Arora)
(ISBN 0-486-24643-4)

15. *Trees of the Northeast Coloring Book*, Stefen Bernath (1979)
(Copyright © 1979 by Dover Publications, Inc.) (ISBN 0-486-23734-6)

16. Miscellaneous

    State Birds and Flowers Coloring Book, Annika Bernhard (1990)
(Copyright © 1990 by Dover Publications, Inc.) (ISBN 0-486-26456-4)

    Long Island Nature Preserves Coloring Book, Sy and Dot Barlowe (1997)
(Copyright © 1997 by Dover Publications, Inc.) (ISBN 0-486-29406-4)

    Botanical Gardens Coloring Book, Dot Barlowe (1997)
(Copyright © 1997 by Dover Publications, Inc.) (ISBN 0-486-29858-2)